By Ken & Fran Rubin

Produced & Designed by Howard Brookstein

Crown Publishers, Inc., New York

Dedication

To all the inventors and artists of yesterday who created these treasures, and to all the people today who enjoy them.

Acknowledgments

Only with considerable assistance from many people could we have presented our favorite subject to the rest of the world. We sincerely extend our thanks and appreciation to:

Renée & Bernard Rubin for writing assistance and encouragement. Howard Brookstein, through whose professional skill and hard work it all came together. Bob Legan, our consulting "Arcadologist" and generous supplier of materials. Richard Bueschel for help in getting our gambling facts straight. Florence & Howard Fertig, who supplied the inspiration with which we began.

And also to all the following people who have contributed in many ways to the creation of this book: Roy Arrington, Sam Bergman, Sandy Bernstein, Larry Birnbaum, Burton Burton, Frank Decaro, Fred Fried, Howard Garrett, Mel Getlan, Shelle & Joel Gilgoff, Helmut Gurdemir, Stan Harris, John Herman, Fred Kaplan, Tim LaGanke, Ed Mosler, Allan Pall, Harold Roger, Bob Rosenberger, Mark Schlesinger, Jim Smith, Orin Yeager.

Typography: Ray Zimmerman

A price guide and rarity supplement and additional information about machines are available directly from the authors. Please address inquiries to:

K&F Publishing Co.
Box 240, GPO
Brooklyn, N.Y. 11202

Printed in the United States of America
Published simultaneously in Canada by General Publishing Company Limited

Library of Congress Cataloging in Publication Data

Rubin, Ken.
Drop coin here.
1. Coin operated machines—Collectors and collecting.
I. Rubin, Fran, joint author. II. Title.
TJ1557.R8 1798 629.8'32 78-12753
ISBN 0-517-53298-0

Table of Contents

Introduction

New ages sneak up on us so imperceptibly that we are often halfway through them before we realize that the old has gone. We adjust to new conditions so easily that each marvel of technology is soon accepted and seems always to have existed. Then, as ordinary objects in daily use disappear or are replaced by newer models, what were taken for granted eventually become "collectibles." Museums now exhibit the relics of yesterday with the same seriousness as they do those of the distant past.

The numerous studies of modern technology notwithstanding, for most of us the process that turns scientific theory into invention and invention into production is too complicated to be grasped in its entirety. Sometimes, therefore, in order to understand why we make things, how we use them, and what we use them for, it is helpful to look backward and focus on a single device. In this book we will turn our lens on the early days of the machine activated by the insertion of a coin.

Today we buy our morning newspaper from a machine, deposit our coins in the highway tollbooth, feed the parking meter, lunch in an automatic cafeteria, make a pay telephone call, pass through a coin-operated turnstile, play a coin-operated game, and wash our clothes in a coin-op laundromat. Coin-controlled devices are ubiquitous robots that magically come to life for our pleasure and convenience. These silent salesmen are twenty-four-hour clerks, swift, efficient, polite to everyone. For every customer, the completion of a transaction with a machine is a gratifying experience. The coin-slot has become a universally recognized symbol and is common in almost every society the world over.

And today the manufacturing, operating, and supplying of coin-operated machines is a multi-billion-dollar industry. Tons of coins are taken in daily and returned to circulation. Our coin-operated appliances are sophisticated, strong, and reliable; the latest ones belong to the new micro-electronic age and are reaching new levels of performance. Because of the ever more complex engineering required, however, production has become the province of research teams and marketing specialists, with design committees answerable to finance departments. Superior as today's machines are, the emphasis on efficiency has lost us the sense of individual achievement evidenced by the unselfconscious artistry of yesterday's commercial creators.

From the beginning of the Industrial Revolution in the nineteenth century until only recently, a "better mousetrap" optimism prevailed in the United States. Yankee ingenuity was greatly admired, and success was within reach of anyone with a new idea or invention. America's heroes were her inspired wizards and tinkerers. By 1885, all the world conceded to America her supremacy in the invention and development of machinery. And indeed, in spite of some inequity in the American economy, the distribution of wealth created by industry resulted in more markets for more goods and even greater prosperity. Masses of people with money to spend were concentrated in cities, and a loud jingling of coins could be heard throughout the country. Inventors who saw the potential profits from machines that would accept these coins in a pleasant way, either for goods, services, or amusement, were soon at work. In 1890, Clement C. Clawson, a New Jersey machine inventor, gave two reasons for the growth of the industry. "Americans," he said, "are improvident and never stop to think that twenty nickels make a dollar.... No other country has a standard coin adapted to the devices, and one that is so popular, so uniform, and so slightly valued." Fortunately, "improvidence" can also be viewed as confidence in America's future in which a coin can be so easily and happily spent by dropping it into one of these fascinating machines.

Beginnings

The concept of operating a device through the use of a coin, seemingly so modern, is actually a very ancient one. More than two thousand years ago in Egypt, a coin-operated machine was invented to dispense a flow of holy water. When a worshipper was about to enter a temple, he sprinkled himself with the lustral water purchased from a vase near the entrance. This vase, a truly automatic vending machine, was described around 200 B.C. by Heron of Alexandria in his *Spiritalia.* The device had a coin slot on the top, a pivoting lever inside on which the coins fell, and a valve that was opened briefly by the lever to permit the flow of holy water. When the coin dropped off the lever, the valve shut. The cost, five drachma, or about a dollar, testifies to the business acumen of the temple's priests. Although we have no evidence, there may have been other such mechanisms in use. Note, however, that the important factors in the success of a coin machine — a repeatedly needed product sold in small enough amounts to keep the container a practical size, a location through which buyers were guaranteed to pass, and a standard coin in circulation — were all present in the example of the water dispenser.

Holy Water Vender

When the Roman Empire in Europe disintegrated into feudal enclaves, its monetary system also dissolved, and for a long time the idea of coin-operated selling machines remained dormant. Then, about 1700, devices for the purpose of vending snuff or tobacco appeared in English taverns. Less sophisticated than the holy water dispenser, they were known as "Honor Boxes" because their profitability depended on the honesty of the customer. In operation, an English halfpenny is deposited and a button is pressed, causing the spring-loaded lid to fly open to allow the customer to withdraw a pinch of snuff or a pipeful of tobacco. The customer was then honor bound to close the lid and pass it along to the next patron. The product, the popular gathering place, and the coin fulfilled the requirements for successful vending machines, but the need to watch for cheats impeded their usefulness.

Honor Box

Although other forms of mechanical coin depositories existed throughout history, such as banks, church poor boxes, fare boxes, cash registers, and coin sorters, they are not considered coin-operated devices because no product or service was sold directly through them. Not until the early nineteenth century was the principle of coin operation readopted for popular use, and for an unusual reason. According to G.R. Schreiber, in his pamphlet *A Concise History of Vending in the U.S.A.*, in 1822 an English bookseller named Richard Carlisle devised a machine to sell controversial literature in the hope that through the anonymity of the selling device he could avoid prosecution from the censors. (The courts, however, still found him guilty.)

In the United States, the first coin-operated machine appeared in 1839. Based on an earlier English patent, it is the oldest coin-controlled device covered by law. Labeled "Penny Papers" (p. 6), it vends a paper-wrapped packet of tobacco. It is operated by inserting an early large American penny and pressing the button, causing the spring-loaded side drawer to slide open. Though similar in design to the "Honor Box," it eliminated the honor system and was thus an improvement. Its flaw, however, is that it requires that the drawer be closed before each sale or the patron loses his money.

The next impetus for the invention and production of vending and service machines came from the increasing availability and sophistication of tools for the mass production of sturdy containers and mechanisms. In Europe the number and variety of coin-operated machines continued to grow throughout the middle eighteen hundreds. An English patent was granted in 1857 for a postage stamp vender, and in 1867 a German inventor built machines to sell handkerchiefs, cigarettes, and confections. In France and other countries, too, inventors were turning their attention to the concept of vending machines.

But success was only moderate. Once machines were designed and built, their profitability depended on the willingness of merchants to buy them and a steady public to patronize them. The necessary ingredients for a new industry, such as the widespread distribution of discretionary money, and other social and economic factors, were apparently not present in the right proportions.

In the United States, however, the end of the Civil War had paved the way for a second industrial revolution based on the internal combustion engine and the electric dynamo. The spectacular growth of American invention and industry and the change from an agricultural society to a capitalistic urban industrial one, laid the foundation for the subsequent expansion of the entire coin-operated machine business.

In America, too, inadvertently at first and deliberately a few years later, a new function for coin machines came into being. On August 15, 1871, a patent was granted to John Hall of Watertown, Massachusetts for a mechanical bank called "Race Course Toy Bank" (p. 7). The toy had a cast-iron base and two tin horses which revolved on top when a nickel was inserted and a string pulled. But it seems that the novelty of the action appealed to more than just the thrifty, because the bank soon found its way into saloons, where those who preferred to speculate rather than save found amusement in betting on the horses. Though not originally intended for gambling purposes, this mechanical bank is now considered the earliest coin-operated gambling device.

Six years later, another mechanical bank was patented with a decidedly intentional gambling aspect (p. 7). Although titled "Guessing Bank," it had little to do with saving. In use the patron called out a number from one to six and deposited his coin in the hat of the seated figure. The coin caused a wire to spin around the dial, and if the wire stopped on the number guessed, the proprietor paid five coins as a reward. The odds, of course, distinctly favored the proprietor. Patented on May 22, 1877, by Edward J. McLoughlin of New York, this bank is the oldest *designed* coin-operated gambling device.

Most of the early machines were regarded as merely ingenious novelties, but between 1884 and 1889 a major industry was born, catalyzed by the influx of foreign inventors who had already patented and produced machines. Most were vending machines, except for German coin-operated scales imported by an American group. During these years, the rapidly rising number of American patent applications testified to the growing importance of the coin-op business.

It must be remembered that although it was the public who bought from the machines, the primary target for manufacturers was the purchaser of the machine itself. The advantage to the operator in a country where labor was in short supply is obvious. Despite problems such as breakdowns, cheats, and vandals, machines were sober and reliable, demanded no wages or holidays, stole no money, joined no unions. Often they sold, at a good markup, products too trivial to be sold conveniently in other ways — a whiff of perfume, a collar button, a box of matches. And they captured money that might otherwise not have been spent.

In the beginning, all coin-operated machines were called slot machines (derived from coin-in-the-slot machines), but the name was eventually applied specifically to the infamous gambling devices that were becoming so popular. By 1891, simple nonpayout devices were appearing in most cities, designed around themes of poker, dice, and roulette. "Taking a chance" by playing a machine seemed to be a harmless and entertaining pastime to the player and a new source of profit for the manufacturer and operator. The rapid spread of these machines was followed by the equally rapid appearance of antigambling crusaders, but the gambling machine industry continued to develop and expand.

Coin-operated amusement vending also found favor with the American public. About 1891, the newly commercialized Edison phonographs with coin-slot attachments were set up in special shops (Edison parlors, p. 8) in high-traffic locations. The success of these parlors inspired operators to seek other types of amusement machines. Kinetoscopes, stereoscopic viewers, strength testers, and fortune tellers were brought together to create the first Penny Arcade.

The 1890s were the years during which the industry developed and expanded. Machines could be found in every town, on telegraph poles and steamboat landings, in saloons and cigar stores, at newsstands and resorts. A gambling device was even advertised as being small enough to take along for picnics and holiday outings. Companies were formed specifically to operate routes of machines. Distributors and jobbers set up to serve the growing market.

Penny Papers

A brass tag near the coin-slot reads: "Manufactured By Green & Broad, New York, Corner Ave. C & 3rd S." A New York City directory lists Green & Broad as locksmiths at this location from 1839 to 1845. The machine is made of painted cast-iron and brass, and advertises "W. H. Nicholas, Extra Tobacco."

Race Course Toy Bank

Creators who also possessed business know-how began to dominate the field. Production went from a local operation to a national enterprise; even exporting overseas became successful. The three major divisions—gambling, vending, and arcade—were established.

Only a few of the old giants in the coin-op industry still survive. Some of the major companies that are familiar names to collectors today are sketched here, but the history of the Mills Novelty Company epitomizes the business in its heyday.

Mills Novelty Company. The most famous name in coin-operated machine history began in 1897 when Herbert S. Mills purchased his father's business and changed its name from the M.B.M. Mfg. Co. (founded 1893) to the Mills Novelty Co. The father, Mortimer, was a mechanical genius, having invented, patented, and produced numerous devices in many fields, including pneumatic railroad gates and cigar venders. Though Herbert was not an inventor, he was an expert businessman, having gained experience in the industry by working for his father and operating and manufacturing machines on his own. Herbert's other brothers subsequently joined the firm and the youngest, Bert, developed into an inventor like his father. Bert was the genius responsible for many of the company's later successful products. Fred Mills, Herbert's son, took over the presidency in the 1920s.

The great success of the Owl floor model slot machine, invented by Mortimer in 1897, quickly established Mills as the leader in the industry, a position the company would maintain for over a half century. Mills adopted the owl motif as the company's trademark (p. 10) and proceeded to create hundreds of other successful designs in every category—gambling, vending, and arcade—including the famous Violano-Virtuoso, an automatic violin-playing machine. Herbert's ability to merchandise his and other firms' products through catalogs, trade exhibitions, world's fairs, and branch offices necessitated a move in 1907 from the original location in Chicago at 11 South Jefferson Street to larger facilities at Jackson Boulevard and Green Street. The new headquarters employed over 600 people, and still stands today.

In its earlier years, Mills also operated its own vending routes and penny arcades, besides selling machines. The boom years before World War I saw their finest machines produced, but the period between World War I and the late 1920s was a lull for the industry, and few exciting new products were marketed. When Mills introduced the modernized three-reel slots in the early 1930s, the company's business once again thrived, despite the economic depression. Through this decade, however, Mills's attempts to diversify into the growing pinball and jukebox business were not successful. During World War II the company's name was changed to Mills Industries, Inc., to present a new image. After the war, Mills resumed production of its gambling machine line, and although they regained their former preëminence in the industry, the market itself was shrinking because of the growing list of states and localities

Guessing Bank

prohibiting gambling. Finally, the 1951 federal law prohibiting interstate shipments of gambling equipment badly wounded Mills and the industry in general. Without their former diversified lines, Mills was effectively out of business.

Through the years Mills Novelty Co. turned out products known for quality of design and workmanship. The company's extraordinary ability to capture a part of American culture in its products over so long a time is evidenced in their many machines that have survived.

Caille Brothers Company (pronounced Cail). In 1901, brothers Adolf and Arthur Caille formed their famous firm in Detroit by merging their respective companies, The Caille-Scheimer Co. (founded 1900) and The Caille Co. (founded 1897). Although second to Mills in size, the Caille Brothers Co. was second to none in styling and craftsmanship. In all three categories, their machines were generally more highly embellished than those of other manufacturers. The company sold out to the Fuller-Johnson Corporation in 1932 who continued the name until the late 1930s.

Watling Manufacturing Company. In 1902, Thomas Watling and his brother John bought the firm they had been working for in Chicago, D.N. Schall & Co., a pioneer slot manufacturer, and changed its name to the Watling Mfg. Co. Over the years, Watling has produced many famous scales, and vending, arcade, and gambling machines. They are still in business today producing scales.

Charles Fey Company. A truly original inventor and tinkerer, Charles Fey started his business in San Francisco in the early 1890s, manufacturing and operating a variety of slot machines and scales. He is most famous for the development of the three-reel slot machine he named the "Liberty Bell." Fey continued manufacturing and operating machines on a limited basis into the early 1930s, when the Depression forced him to close.

Clawson Manufacturing Company. Clement C. Clawson, another prolific pioneer of coin-operated machines, incorporated his company in 1891 in Flagtown, N.J. He invented and produced many dozens of unusual gambling and vending machines and had many patents in his name. The company is still in business today manufacturing ice equipment in Boonton, N.J.

Exhibit Supply Company. This company was founded in Chicago in 1902 by J. Frank Meyers. It was originally a printing firm that supplied fortune cards to machine manufacturers. In the early 1920s, they began producing their own line of trade stimulators, and vending and arcade machines, and by the 1930s Exhibit Supply was the major supplier of complete penny arcades. They were especially known for their line of "claw" and "digger" machines. Around 1960, they discontinued the machine division but continued printing some of the same cards with which they began, and are still in business today.

O.D. Jennings & Company. The Industry Novelty Company was founded in 1907 in Chicago by O.D. Jennings, a former employee of the Mills Novelty Co. In 1920 he changed the name to his own. His line included many famous vending and slot machines, and scales. In the 1930s, Jennings emerged as the major competitor to Mills in the slot machine business, and continued producing slots into the early 1960s.

Rock-Ola Manufacturing Company. Dave Rockola began his company in Chicago around 1927. His many products included add-on jackpot fronts for slot machines, pinball, gambling, and vending machines, and jukeboxes. The company is still in business producing jukeboxes and vending equipment.

The Edison Kinetoscope and Phonograph parlors of the 1890s were the forerunners of the Penny Arcade.

Pace Manufacturing Company. Founded in Chicago in 1927 by Ed Pace, the Pace Mfg. Co. began producing three-reel slot machines, and pinball and countertop games. In 1934 the company introduced the famous "Paces Races" console horse race machine. The company folded around 1951.

Bally Manufacturing Corporation. The unchallenged leader in the coin-op machine industry today began in 1931 when its founder, Ray Moloney, entered the developing pinball machine business with "Bally Hoo." The immediate success of his first product led to other types of coin-op machines. By adding a payout mechanism to a pinball machine in 1933, Bally also entered the gambling industry, which eventually led to developing a line of slot machines. Although the 1951 antigambling laws forced them to discontinue gambling equipment, they continued in business with their successful line of pinball and arcade machines. When the laws were amended in 1963, Bally reentered slot machine production with a revolutionary design called "Money Honey"—the consummate sleek, illuminated, glass and chrome, electronic slot machine. It is this machine that made obsolete all the earlier mechanical slots that we now consider antiques.

Decorative Art—Machine Design

The first consumer goods were barely off the production line when the first outcry against vulgar popular taste was sounded by the critics. The middle and lower classes, they complained, favored garish imitations and stereotypes, and had no respect for consistency of style. Borrowing from other cultures was, and is, a common practice in art, but indiscriminate borrowing for an ignorant public was more than the aesthetic elite could bear. And so began the still continuing war between mass taste and artistic sensibility, applied and fine art, machine and hand crafts.

Profit motive or not, the exuberant eclecticism that characterized the middle Victorian period ultimately stemmed from pride in the ability to reproduce cheaply and in quantity what had hitherto been made painstakingly for the wealthy few. In the sincere belief that a copy or adaptation of what was considered beautiful is still beautiful, manufacturers made wares in Victorian Gothic, Victorian Renaissance, and other styles inspired by earlier genres. The most favored was Victorian Rococo, or Louis XV, although toward the end of the period the straighter lines and simple incised decorations of Eastlake were also employed. At the turn of the century the sinuous curves of a wholly new style, art nouveau, appeared in industrial design, and in the twenties the geometric lines of art deco set a different standard of modern elegance.

Coin-machine makers freely chose design elements from all sources; whatever caught the public's imagination was adopted to catch the public's eye and money. Between 1885 and 1893 most coin-operated machines were rather primitive in design and mechanism, made by amateur inventors, not professionals. As later devices grew more sophisticated and durable, however, competition shifted to novelty and case design. Because slots took the place of a human salesman, barker, shoeshine boy, etc., cases had to do more than hold and protect the mechanism. They had to present a personality to which the patron could relate. Machines were given titles and themes based on well-known current events, folklore, mythology, patriotic sentiment, sports, and other subjects guaranteed to evoke a characteristic emotional response. Leading firms like Mills and Caille, who made coin-in-the-slot machines for all kinds of needs, were able to combine and refine the various elements in their lines to eventually evolve specific looks for different coin-op functions. Gambling devices retained their opulent looks, but a more showy carnival style developed for most arcade and vending machines, although some are delicate and graceful. In all cases, experience and feedback led to further experiment and improvement.

We are beginning now to reappraise the quality of the nineteenth- and early-twentieth-century examples of industrial art, and to reconcile the fine arts to science and technology. The definition of "work of art" has been broadened; the marketplace is no longer viewed as being inherently opposed to artistic merit. True, the machines in this book were made primarily to make money for the manufacturer and operator, but in order to do so, they had to be economical to produce and run, sturdy and attractive, and suited to the uses for which they were designed. Manufacturers made superb use of the abundant materials—cast-iron, wood, aluminum, electroplated nickel—made available by new methods of production. The expert craftsmanship of painters, cabinetmakers, and other skilled technicians testifies to the care taken in executing a design. These antique coin-operated machines are preserved reflections of the visions and taste of a time and place. They were made to communicate, to satisfy the psychological need for amusement, excitement, wonder, competition, and achievement. They still do.

A Collector's View

The hobby of collecting antique coin-operated machines has grown as more people recognize the exciting qualities these machines possess. As an art form, they cover the period from 1885 to about 1950. Within this period, the diversity of style and function is extraordinary. The charm, beauty, and character found in these objects as works of art would in themselves be enough to make them treasures. But antique coin machines offer a great deal more. They are born entertainers, made to delight the spirit of anyone whose attention they attract. They accept our coins and perform as willingly today as when they first appeared. They elicit a sense of nostalgia that conjures up memories of naïve days and simple pleasures. They invite the attention of the craftsman and mechanic to keep them in presentable and operating condition. They are unique artifacts of America's cultural past.

Where to find them today

Coin-op machines were sold to a small commercial market, not to the general public. Few ended up being stored away in attics or basements of homes. Some of the places they may still be found "in the woodwork" are listed below, along with the more active antique and collector sources.

1. Old cafés and restaurants, pharmacies, general stores, ice cream parlors, billiard halls, confectionery shops, hotels and resorts, cigar stores, barber shops, and saloons.
2. Old arcades and amusement parks.
3. Fraternal organizations and clubhouses.
4. Operators and distributors of amusement and vending machines.
5. Antique shops and dealers.
6. Flea markets, antique shows, and auctions.
7. Advertisements in antique collectors' newspapers and magazines.
8. Other collectors.

For the most part, modern, and therefore less desirable, copies of antique machines are not a serious problem confronting the buyer. The cost of tooling and labor necessary to remanufacture a machine today is usually too expensive to be worthwhile. As with many other antiques, collectibles, and works of art, however, when prices become high enough to make producing fakes a profitable enterprise, it is the inexperienced buyer who is most often deceived. This is beginning to occur with coin-op machines of all types. It is quite acceptable to remake parts in order to restore a machine to presentable and operating condition, but making copies of entire machines, or attachments, or combining a little old with a lot of new and passing it off as an antique is forgery and should be exposed.

Areas of specialization

The wonderful variety of coin-operated machines allows the collector to personalize his approach in assembling his collection. A collection can represent all coin-op categories or just one particular aspect. Some of the possible approaches are:

- By general type: Gambling, vending, arcade, and music (another popular coin-operated machine collectible).
- By specific type in each category: In gambling, such as uprights, three-reel slots, trade stimulators. In vending, such as gumball, matches, cigars. In arcade, such as strength testers, electricity machines, card viewers. In music, such as music boxes (cylinder type and disc type), phonographs (Edison style and disc records), nickelodeons (automatic pianos, violins, band organs, etc.), jukeboxes (78 rpm record players).
- By size: Countertop machines vs. floor models (depending on space available).
- By type of materials: The principal material that the machine is constructed of, such as wood, glass, cast-iron, aluminum.
- By manufacturer: Mills, Caille, or your own favorite.
- By styles: Victorian, art deco, primitive, figural, etc.
- By model variations: During the course of production, especially with machines produced over many years, both subtle and obvious changes in design commonly occurred, creating a collectible variety from the same basic model. For example, operating features were improved for practical reasons, and materials and colors were altered for economic and aesthetic reasons.

Rarity

No other Americana collectible has faced such formidable odds against survival as the early coin-operated machine. Although some thousands of old machines still exist, it is certain that the number is but a small fraction of what originally was produced. Attrition in part was due to ordinary wear and tear, and the fragility of construction materials such as brittle cast-iron, glass, or delicate papier-mâché. The vulnerability of the machines to vandalism, and in the case of gambling machines, deliberate destruction, was also not conducive to survival for the great majority. For the very reasons that the machines were often so short lived, the manufacturers kept improving designs and workmanship, thereby creating a remarkable variety of products.

The production quantities of each machine varied from a few handmade pieces to tens of thousands, depending upon the machine's practical success and its manufacturer's ability to produce and sell it. No company production records are available, though sometimes serial numbers on the machines themselves can provide clues to the number of that type made. By and large, the more produced, the greater the likelihood that more have survived. Exactly how rare any machine is today cannot be stated with certainty since many people own machines outside the group of active collectors who are the compilers of such statistics. The rarity of the machines pictured in this book ranges from one-of-a-kind to hundreds-of-a-kind known amongst collectors. In general, the earliest machines are the rarest ones, with few surviving that were made during the experimental years prior to 1890. A small percentage remain from the period from 1890 to 1900. when the industry developed into its major segments. A large quantity survive that were produced between 1900 and 1915, when the major companies became established, and the bulk that is left that was made between 1915 and 1950, when the industry reached maturity.

The purposeful destruction of gambling machines by governmental authorities began shortly after the machines were introduced. Crusades by reformers to rid their localities of the evil "nickel-slots" frequently resulted first in local and then state and federal laws banning their usage. It was often a race between the manufacturers to design and supply machines to a territory and the enforcement agencies to confiscate and destroy them. Fortunately for the collector, certain factors have increased the survival rate of antique gambling machines: The potential for making money kept many owners from discarding them, or gave the owners an opportunity to resell them. Very large quantities were usually produced, particularly in the case of three-reel slots made from the 1920s on.

Vending machines, in contrast, were of no value once their useful life was over. A change in the product's selling price, package size, or discontinuance of the company or its product altogether, made the singularly designed vender worthless. For this reason, store owners did not often buy the venders but received them on loan from the manufacturers when they purchased the merchandise that was sold in the machine. Sometimes the store owner had an arrangement with an outside operator with whom he shared the profits. Consequently, obsolete machines were returned to the manufacturer, distributor, or operator, who most often junked them. Other factors as well account for making venders scarce. Machines that vended salted nuts would corrode from the salt, and outdoor machines were good for only a few years' operation. Many venders were particularly fragile, having delicate mechanisms and cases made of glass. Because these 24-hour clerks were usually left unattended, vending machines were more vulnerable to vandalism. On the other hand, venders that sold staple products, such as gumballs, cigars, and matches, were made in large quantities. Proprietors were also more willing to purchase these types outright, and later, to store them in the basement, rather than intentionally destroy them. Thus, more of these kinds have survived.

Arcade machines suffered most from daily use, especially from children, and had to be sturdily built. Their gradual destruction was anticipated by their owners, who kept them in operation for as long as possible, and patched and repatched them with parts taken from broken machines. Strength testers literally took beatings and brutal punishment. Lung testers were outlawed and subsequently discarded because of their unhealthy tendency to spread disease. Amusement parks maintained dumps where obsolete machines were set afire. Waterside parks merely dumped the unwanted machines into the drink.

Essentially, machines that could no longer earn a profit and justify occupying commercial space were simply thrown away. Any machine not disposed of still faced destruction by rust, rot, floods, fires, and even war drives for scrap iron. The machines are now collected for their beauty, usefulness, and historical value. Their rarity provides an additional challenge.

About the Photographs

The photographs shown have been taken by the authors over the course of many years while visiting collectors around the country. With an accent on the turn-of-the-century, the pictures cover the finest examples of the art form, from the rarest machines to the most popular, in both original and restored condition. They have been selected to demonstrate the wide range of styles and designs in each category, and within each category they are grouped to show the diversity of imagination used to achieve a similar result.

Naturally, many machines could not be included, for there are still many more splendid examples that have somehow managed to survive. And between original designs and production variations, "new" old machines are constantly being discovered. It is also not possible to include photographs of the wonderful mechanisms that are the raison d'être for the exterior design. Some mechanisms are complicated or delicate, some crude or awkward. Some are preposterous Rube Goldbergs and some are masterpieces. All are marvelous contrivances made by geniuses, eccentrics, and schemers.

Since it is not feasible to describe each machine in detail, here are some points to keep in mind that can be compared visually from one machine to another:

Variety of coin-slots. The coin-slot is the only item common to every machine, but how it is designed is quite varied. Some are separate parts, some are incorporated into larger castings. There are those that include instructions and others that indicate price. They are shaped as circles, slits, slides, and chutes, and are located in different places on each machine.

Coin denomination. Most machines are penny or nickel operated. Higher denomination machines are mainly for gambling, some of which also accept various coin denominations or trade checks.

Method of operation. In all but the fully automatic (clockwork and electric motor driven) machines, the patron would deposit, lay, place, insert, or drop a coin into the slot and then turn, press, pull, grasp, or push . . . a crank, knob, rod, handle, or plunger. Each machine has its own manner of delivering the reward, merchandise, service, or surprise.

Marquees. These are the sign holders or decorative tops that surmount a machine. Usually they contain printed information that announces the attraction, coin denomination, instructions, or merchandise. They were made in every material and were popularly used in the early 1900s, particularly on arcade machines. Because they were separate attachments, they were more subject to damage and often broke off entirely and disappeared.

Payout cups, vending chutes, scoring dials. Where does the machine deliver its reward or product, or register a score? Note the variety of shapes and locations.

Material and techniques of manufacture. The large variety of material used for construction and the numerous processes used in production are primary reasons for the visual appeal and diversity of the antique machines. As new materials became available in the market and were priced economically, manufacturers incorporated them into their designs. Today some of these materials are no longer available, and some of the techniques are too labor intensive, requiring skills not readily found.

- Metals: Metals are used for both construction and ornamentation. The most frequently employed were iron, sheet steel, and aluminum, with brass, tin, and copper to a lesser degree. Some of the manufacturing techniques used were casting, stamping, embossing, spinning, and machining.
- Wood: Wood was the basic construction material because of its low cost, ease of fabrication, and beautiful appearance. Some very early machines even had mechanisms made of wood. The most popular wood was oak, and up to the 1920s the fancier grained quarter-sawed cuts (tiger oak) were used. As furniture styles changed, and availability of fancy oak decreased and prices increased, the straight-grain cuts were substituted. In a few instances, machines were made of mahogany, cherry, pine, poplar, and walnut. Veneers, too, were utilized. Old World cabinetmaking, hand carving, and lathe turning are some of the special examples of fine workmanship to be noted.
- Glass: Few machines have no glass at all. Glass was mirrored, beveled, molded, curved, and used as a magnifier. Reverse painting, glue chipping, etching, and decaling were methods of decoration.
- Porcelain: Because of its durability, porcelain was an important material for many years. Its weatherability was especially useful for vending machines and scales in outdoor locations.
- Papier-mâché: By far the leader in making papier-mâché compositions was the Mills Novelty Company, which produced many unusual arcade machines in this material. Exhibit Supply also used it for arcade machines but to a lesser extent. After molding, the papier-mâché was hand painted.
- Decals: Decals are decorative or convey information. Paper and water decals were often affixed to glass. Varnish decals were applied to wood.
- Electroplating: The flashiness of plated metal made it a popular finish for any type of machine. The common finishes are polished nickel and copper, and antiqued copper. Special finishes were a striped antiqued copper and "Roman Gold" (an imitation gold).
- Painting: Most enameled metal and many signs in marquees were hand painted or lettered. A special effect used to achieve a rich appearance was the imitation marble and onyx finishes applied over wood or metal on gambling and arcade machines. Mills and Caille were particularly adept at this technique.

To a lesser extent, paper, rubber, leather, horsehair, cloth, gesso, and plastic (Bakelite and Celluloid) were used in all types of machines.

While every effort has been made to provide accurate information throughout this book, due to the sheer number of variables involved complete accuracy may not be achieved. Sources that have been researched are machine authorities, old-timers, collectors, original manufacturers' sales catalogs and advertisements, patents, trade publications, and the machines themselves. With regard to the captions under the photographs, some points of information are further explained:

Names: Most machines are named by the manufacturer. Names are found on the machines themselves, or obtained from catalogs. Some are popular names given by collectors and some are derived from the product vended. Occasionally a name is abbreviated where space is limited. Often names are amusing, taken from inventors, products, and companies, animals, heroes, and myths, and cities, sports, and patriotic themes.

Manufacturers: Many machines were designed, patented, and produced by their inventors. Sometimes the inventor sold his design to a manufacturer or received royalties for its use. Manufacturers sold their machines directly to locations or through distributors or jobbers. Some only leased their machines or operated them themselves. Where the name of a manufacturer is not known, the name of a distributor or operator is used. The same machine may have been made by more than one company, due to normal business changes, such as a company being bought out by another or a merger.

Dates: The dates listed are those estimated to be the first year the machine was introduced for sale. However, a machine may have taken many years to have been conceived, designed, and tooled. And a machine could have been successful enough to be produced for many years, so that the one pictured could be later than the date specified. A *circa* denotes less certainty.

Patent Dates: Patent dates are stated when known to apply directly to the described machine. Since it usually took one or two years to receive a patent after filing an application, the machine may be a year or two older. But patent dates are not always reliable. Some machines list patents indiscriminately, seemingly to discourage potential imitators. Occasionally the dates are entirely fictitious. Often a machine lists a patent that it borrowed a specific design from, in addition to its own new overall patent. When there is more than one applicable patent, the earliest is used.

In essence, every antique coin-operated device is a gambling, a vending, and an arcade machine. As for gambling, every customer is betting that the machine will operate after depositing his coin. (Who hasn't lost his money in a coin-operated machine?) As a vending machine, it dispenses a portion of goods, services, or entertainment. And as an arcade machine, it is an attraction to the eye and a temporary diversion. They are all designed to lure a customer and empty his pockets. Each machine tells a story through its style, its operation, and its history.

Picture yourself standing before each machine and depositing your coin. Imagine the wheels spinning round and winning a jackpot. Imagine purchasing a gold collar button, filling your lighter, shining your shoes. Imagine testing your strength, hearing a tune, seeing a show. It's all guaranteed to raise a smile.

GAMBLING

THE THREE JACK POT.

We make these Machines for either Nickel or Penny.

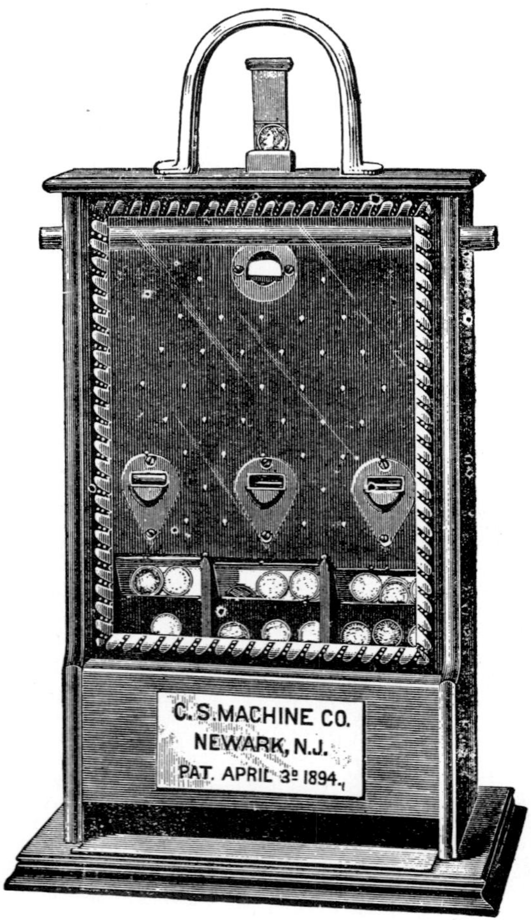

THIS machine is the most attractive and amusing as well as the most profitable of all slot machines.

To load the machine, pull the bar running through the top of machine out to the right as far as it will come. In the back of the bar is a recess. Put a nickel in this recess and push the bar back as far as possible, and repeat until you have enough nickels in the right-hand side pocket; then load the left-hand pocket with the left-hand end of the bar. The two side pockets only are to be loaded.

To operate machine, put nickel in recess in top and front of push rod; then press rod all the way down and the nickel will find its way down between the pins into one of the pockets, or will go into one of the three slots, in which case, push the rod down again and it will open the pocket under the slot into which the nickel went. This nickel is retained in the machine and therefore registers the number of times the pockets are opened.

Made only by the **CLAWSON MACHINE CO.,**

To whom all correspondence should be addressed for prices and information.

21 Hackett Street,

Newark, N. J., U. S. A.

Gambling, from casual betting to ruinous obsession, has been analyzed by novelists, mathematicians, psychologists, theologians, and just about everyone else. So far, there has been no general agreement concerning the causes or the morality of gambling, but there is no doubt at all that it is a universal and deeply ingrained human activity. From the recognition that life is shaped at least in part by unforeseen and uncontrollable events (chance) comes the effort to know whether or not those events will be favorable (good luck) or harmful (bad luck). In modern times, divination is no longer regarded as valid, but staking a possession on the possibility of getting back more than the stake (good luck) endures.

Almost anything for which the result is not predictable can be bet on. Throughout history, cards, wheels, dice, and coins were employed to produce purely random results. Inevitably, machines were invented to replace the human card dealer and dice tosser, and to simulate sporting events like horse racing. Because the actions involved were few and discrete, it was a simple matter, by the late 1880s, to adapt these machines to coin-operation. Not long after, other popular gambling games using wheels and coins were also mechanized and made coin-operated.

These early machines, though popular, were nonpayout and relatively simple to produce, intended mainly to stimulate trade. Then, in 1892, Jones Novelty Company introduced one of the earliest payout machines, the Three For One (p. 28), which was an immediate success. The player was thrilled by the fall of coins (if he made a hit), and the operator was relieved of the need to check the machine and pay the winner.

Other companies were quick to copy the automatic payout feature and certain improvements, such as the visible jackpot of the "Jacks" machines (p. 29), were copied in turn. The color pinwheel, an upright variation of roulette, was even more popular than the coin-drop, possibly because of the hypnotic effect of the colorful spinning wheel. Like the coin-drops, the first of these wheels were countertop models, but their commercial possibilities inspired the production of elaborate upright cabinet floor models.

Again, competition led to more improvements, sturdier and more sophisticated mechanisms, and new ways to tempt the player. In the 1890s, some successful companies were: Automatic Machine Tool Co., Berger Manufacturing Co., Clawson Machine Co., Cowper Machine Co., Charles Fey & Co., Charles T. Maley Co., Paupa & Hochreim, Daniel N. Schall Co., Victor Novelty Works, Waddell Wooden Ware Works, and White Manufacturing Co. As the markets grew, the need for more capital and greater production and distribution capacity resulted in the creation of larger firms such as Mills Novelty Co., Caille Bros. Co., and Watling Manufacturing Co.

The first major change in slot machine design came in 1905, when Charles Fey, after years of experiment in his San Francisco workshop, produced the first three-reeler. Based on the single-reel trade stimulator, the "Liberty Bell" (p. 33) has a three-reel combination and payout mechanism. Symbols are horseshoes, bells, stars, and playing-card suits. In operation, a perforated plate spins with each reel. On a winning combination, the perforations line up so that metal fingers can protrude to trip the coin-slides for a payout.

Because of his limited production capacity, Fey brought his invention to Herbert Mills to manufacture and distribute. Mills had the machine redesigned to make it more reliable and cheatproof, and called the new 1909 model the Mills "Liberty Bell." For his gum-vender model, Mills produced his own Liberty Bell Fruit Gum, and changed the card symbols to fruit symbols, which he copyrighted. This variation was called the "Operators Bell" (p. 33). Both operators and patrons appreciated the faster action of the three-reeler over the slower choose-your-color pinwheel machines.

Patents for gambling mechanisms were no longer obtainable, so that successful models were often copied part for part. Other manufacturers such as Caille Brothers, Industry Novelty, and Watling Manufacturing quickly brought out their own three-reelers. Although other machines, such as the spinning arrows (p. 32) and the double-dials (p. 31), were still being introduced, the growing popularity of the three-reeler made the others obsolete.

For the next few years, except for minor improvements, the three-reeler did not change dramatically. But in 1927, the industry was perked up by the reintroduction of the jackpot, which had been abandoned along with the upright one-wheelers and "jacks" machines. In 1931, Mills Novelty Co. introduced their "Silent Jackpot Bell" (p. 39), which began a new era in slot design. Inside, the mechanism was redesigned for less noise and less muscle to pull the handle. A new horizontal escalator added visual interest and better slug detection. Outside, the traditional slot look was changed to a brightly enameled, flashy appearance.

By now, machines were everywhere. In spite of the Depression, trade stimulators and payout machines were more popular than ever, as the public found them to be a cheap and sometimes rewarding form of entertainment. Besides Mills, Caille, and Watling, other big names in the industry enjoying success at this time were O.D. Jennings & Co., Pace Manufacturing Co., and Rock-Ola Manufacturing Co.

In all this time, while not wholly approving gambling devices, the public was not entirely disposed against them. But when organized crime, spawned during Prohibition, muscled in on the industry during the Depression, local officials passed laws to ban gambling machines entirely. The market began to dry up. Production was further eroded by World War II, and afterwards a more limited manufacture was resumed. But the most crippling blow came in 1951 when Congress made it a federal offense to transport slot machines, consoles, or even their components across state lines. Shortly afterwards, Illinois passed a law prohibiting all out-of-state shipments, and since Chicago was the center of the gambling machine industry, this was the end.

Until 1951, even with legal constraints, the money-making potential for the industry was enormous. New restrictions were met by clever evasions — innocent tokens were substituted for evil money, machines were made small enough to be quickly hidden away, or ostensibly to sell music, gum, or fortunes, or to test skill. In less prejudiced areas, gambling purposes could be more frankly shown by machines such as the large and opulent uprights and those with visible jackpot windows. And patriotic names or themes, such as the "American Beauty" or "Liberty Bell," added further protective respectability.

In America, as in all societies, gambling has always been regarded as immoral, and the greed and deception associated with the coin-operated machine further justified the attitudes of antigambling forces. Machines were manipulated by both sides. If unhappy with the built-in odds, the greedier operator could use special tricks: He could move or add nails on the coin-drops to make winning even more unlikely. Holes on one-wheelers and three-reelers could be plugged. Symbols could be changed or a "bug" (a small metal half-circle) could be fitted over the cog to prevent a payout. The popular term "one-armed bandit" was not far from the truth. The patron could be just as devious: His use of magnets, wires, drills, and slugs perpetuated the battle with the manufacturer who continued to invent new methods to combat the cheating, such as gooseneck coin-slots and slug-detector windows. Around 1946, inveterate gamblers even learned to memorize the position of the reel symbols and time the pull of the handle to beat the three-reelers. This "rhythm system" was countered by the manufacturer's introduction of a random timing device.

In recent times, the trend against gambling has been reversing. The legalization and sponsorship by state and local governments of assorted forms of gambling has helped to remove some of the stigma surrounding gambling machines. Still, it is necessary here to point out that there are federal and state laws that prohibit transportation and possession of gambling devices. This includes all the types of machines pictured in this section. The authors and publisher of *Drop Coin Here* in no way advocate the breaking of laws concerning gambling devices. It is unfortunate that these precious relics of Americana can be confiscated and destroyed in order to conform to a law passed in a different era.

Some state legislatures, however, have recently been enlightened by collectors and have amended their restrictions to allow for private ownership of antique gambling machines not intended for illicit use. (Most of these laws define antique gambling devices as those machines manufactured prior to January 1, 1941, and others set an age limit of twenty-five years or older. For the collector the latter is more preferable since it allows for mostly all the mechanical machines that were produced after World War II that still fall into the antique category.) As of this writing, the states that have made these changes are: California, Florida, Minnesota, Missouri, Nevada, New Hampshire, Oregon, Pennsylvania, South Dakota, Texas, Washington, and Wisconsin. Bills are still pending in many more states, and the reader would be wise to check with state authorities for specific information.

Once a person has decided to collect these fascinating devices, he will encounter as broad a range of prices as types of gambling machines. In good to excellent condition — either restored or original — a single upright will generally range from \$3500 to \$10,000. A "Jacks" machine could be picked up for between \$250 and \$1500, whereas the more popular three-reelers command from \$350 to \$1500. Expect to pay \$1000 to \$3500 for a trade stimulator with a token payout, and between \$500 and \$2000 for a trade stimulator.

A pair of slot machines with side venders are operated in this 1920s restaurant.

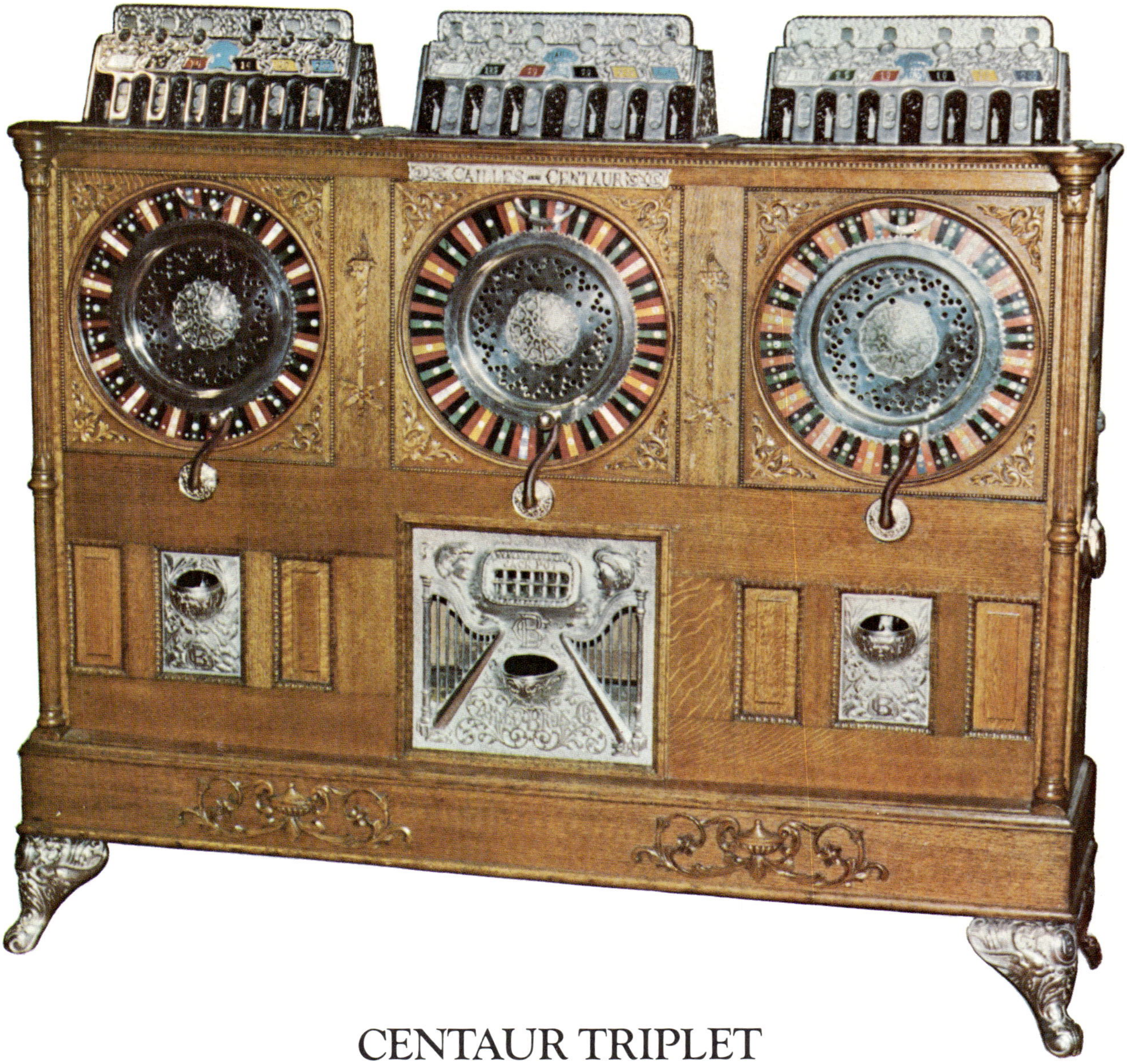

CENTAUR TRIPLET

Only the most elegant saloons and casinos could offer their patrons such a splendid gambling device. The massive hand-carved quartered oak cabinet houses three individual Centaur mechanisms: two nickel machines flank the quarter one with its alluring coin-filled jackpot window. The action starts when a coin is dropped in one or all eighteen slots, each slot's color corresponding to one on the wheel. (Less frequent colors pay higher odds.) Pushing down the lever on the coin head registers the bet, and pushing the front handle releases the notched wheel to spin. The rotation stops when the arrow drops back into a notch. If the arrow stops at the color bet on, the Triplet automatically pays out. Because of its bulk and high cost, few manufacturers offered triple cabinet slot machines. This example was made by the Caille Brothers Company in 1905.

BIG SIX — LONE STAR TWIN

This magnificent double, lavishly embellished with copper-plated iron castings, houses two of Caille Brothers' most attractive uprights. The Big Six features a colorful etched mirrored glass with Cupid on the moon aiming his arrow at the lovely goddess. The Lone Star is unusual in that the arrow, not the wheel, spins when the handle is pressed. In order to circumvent local gambling laws, the music box (framed in the lower casting) automatically plays tune each time machine is operated. 1904.

DEWEY

America's pride in the victorious naval hero of the Spanish-American War prompted Mills Novelty Co. to produce this patriotic Dewey upright. Portraits of the admiral are in a front casting and painted on the medallion of the etched mirrored glass. Operators appreciated the machine's reliability and large earning capacity, and the Dewey continued to attract the customers from its inception in 1899 until around 1930, making it the longest produced upright. As a result, more of these have survived than any other upright slot machine.

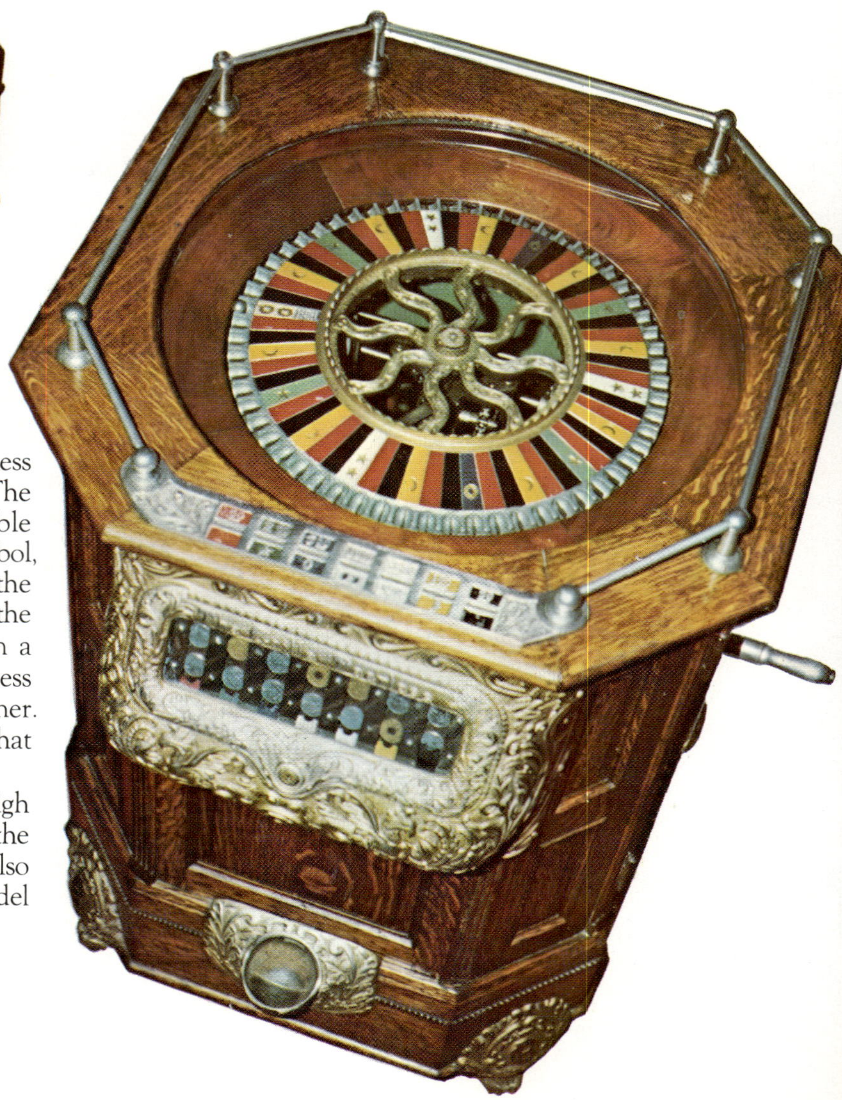

PEERLESS ROULETTE

Advertised as a "slot machine par excellence," this Peerless model was introduced by Caille Brothers in 1907. The croupier has been efficiently replaced by a remarkable mechanization. The gambler selects his lucky color or symbol, the odds for which are cast into each coin slot. When the handle is cranked, the roulette wheel revolves, spinning the metal ball around the rim until the ball drops back in a pocket on the wheel. When the movement stops, the Peerless will automatically pay out the designated reward for a winner. By operating the roulette wheel himself, the player feels that he can control his own luck.

This handsome and sturdy gambling device stands 47" high and was available in handcrafted mahogany as well as the quartered oak pictured here. Mills Novelty Co. was also manufacturing a similar (although less elaborate) floor model roulette at the same time.

ECLIPSE

Caille Brothers manufactured many upright slot machines in this Victorian rococo styling, such as the Big Six, Centaur, Special, and the above Eclipse. Like most uprights, they are all over five feet tall. Castings and wheels were changed or retained and mirrored glasses added or eliminated to give variety to what is essentially the same machine. Here, the lower casting has a romantic scene of Cupid kissing a damsel while cherubs and lion paws form the cabriole legs. 1905.

VICTOR

While Caille and Mills dominated the field, other companies also produced attractive and successful uprights such as this musical one-wheeler made by the Victor Novelty Works in 1906. The oxidized-copper castings and wooden hand-carvings are as eye-catching as any of those put out by the better known rival companies. The brightly colored glass sporting the Victor logo remains stationary as the wheel behind it spins.

20TH CENTURY

The name is symbolic of the advent of the new century and the implication of a new era of progress and excitement. Mills Novelty Co. enthusiastically merchandised this one-wheeler by offering the operator a variety of alternative features. Different heads could accommodate nickels, quarters, half or silver dollars. And for sensitive locations, trade check or musical models were also available. All versions have the unusual side crank handle. Made in 1902.

VICTOR

The unique sloped front cabinet makes the Caille Victor a real standout. Caille Brothers repeated this distinctive shape in its Venus and Quaker uprights, all three of which are rare machines today. The bright blue and yellow embossed and lithographed tin wheel shows a ring of narcissistic Pucks perched on daisies. A lively tune plays while the wheel spins to guarantee that win or lose, each player receives some enjoyment for his nickel. Made in 1905.

NEW CENTURY DETROIT

Holes in the metal wheel allow the patron to see that the dial is not plugged, reassuring him that any color could win. Called the "Honest John" payout when first introduced on the Detroit, this feature was frequently used on many later Caille uprights. However, an unseen internal percentage regulator allows the operator to continue to control the odds and rake in a 10 to 75% profit at his own discretion. Made by Caille Brothers Co. in 1901.

HULL'S CHICAGO

Automatic Machine Tool Co. offers a less expensive upright by substituting decals for castings. The overall effect, however, is quite attractive. The operating technique does not vary much from other upright slots. The player still drops a coin in one of the six coin slots that correspond to colors on the wheel. Pulling the handle on the coin head registers his bet and keeps the last coin played visible in the window to allow for slug detection. Flags pop up above the dial to indicate the color bet on. Made in 1904.

BLACK CAT

Music, the beveled mirror reflecting light, and the glint of the nickel-plated castings make this upright a shining example of its kind and hard to pass by. Instead of the standard payoff, additional enticement is offered by visible jackpots. Losing coins drop randomly into the six windows. The player wins whatever amount is in the corresponding jackpot when he hits, whether it holds its maximum of $10 or nothing at all. Caille Brothers Co. 1902.

OWL

The Mills Owl, the first mechanical upright cabinet slot machine, proved more reliable and convenient than the earlier battery-operated electric models. Unique to this upright is a bell that rings to announce a winning payout. A five-way head sits atop the beautifully carved oak cabinet, and a circle of owls are perched on the lithographed tin wheel. Because of its immense success, Mills Novelty Co. later adopted the owl motif as its trademark. 1897.

BUFFALO

Once the overall styling of the upright slot machine was established, it did not vary greatly from one company to another. Watling Mfg. Co., essentially a copier, produced its own line with mechanisms almost identical to those of the leaders in slightly varied cabinet designs. Naming this upright after the popular symbol of the Western frontier lends even more distinction than its mirrored glass and refined castings do. Because they produced a smaller line, few Watling machines exist today. 1903.

LION

Here is yet another animal in the upright menagerie. The five-way head accommodates individual bets on each or all of the four colors (paying two, five, and ten coins for red and black, green, or yellow) and on the King of Beasts for the big jackpot of twenty coins. By 1910, upright cabinets were simplified in order to compete with the increasing numbers of lower-priced counter slot machines. Caille Brothers Co.

AMERICAN BEAUTY

"Twelve can play and two can win." The coin receiver allows two bets on each of the six colors. This machine is for the serious gambler looking for more action, and the wily operator looking for more profit. The player can double his bet and the operator can double his intake on each play. The stately oak cabinet is adorned with gilt decals and demonic castings, and stands almost six feet tall. Made by Paul E. Berger Mfg. Co. in 1901.

CRICKET

You'll need a bit of skill to play this upright. Drop five cents in the head and pull the handle to bring the coin into position, and pull the plunger to kick it onto the play field. The nickel ricochets off the pins down the front of the machine, occasionally falling into one of the slots that releases the corresponding jackpot. You can see the money in each cup, but you've got to be lucky to hit a full one. A confident player might aim for the left target to release a $5 token. Mills Novelty. Co. 1904.

CHICAGO RIDGE

The earliest floor model slots were battery operated, so when Paul E. Berger Mfg. Co. introduced this model in 1897 it naturally employed electrical contacts to trigger the payout mechanism. However, a disadvantage is that the player must hold the handle down to maintain contact when he hits a winning combination or he loses his reward. Also, the operator had to incur the expense and inconvenience of replacing batteries. When the all-mechanical uprights were introduced, these became obsolete.

DUPLEX

Its size, imposing appearance, and majesty make this one of the most spectacular uprights. Mills Novelty Co. incorporates all of the finest artistic elements in the wood carving, ornate castings, and lithographed tin wheel. In operation, double five-way heads offer a variety of up to ten bets. Corresponding colors identify the jackpot windows; visible rewards show the player exactly what he might win. 1901.

PYRAMID BANKER

If a nickel dropped from the top lands in a winning slot, pushing a knob opens the door for the reward. Nonwinning nickels help to fill the jackpots or fall in back for the operator's profit. Made by Charles T. Maley Co. in 1894.

FORTUNA

You'll need Lady Fortune's blessings with this elaborate nickel drop. Not only must the coin fall into the proper slot, but it must do so when one of the revolving clockwork driven jackpots is directly underneath. Charles T. Maley Co. 1894.

THREE-FOR-ONE

Automatic Machine Co. houses an early payout device in a decorative Eastlake case. In 1892, little else was needed to attract the gambler's nickel. If it dropped into the hidden slot, three nickels were paid.

THREE JACKPOT

A nickel is played by pushing the plunger. If the coin drops into one of the three visible jackpot slots, the patron is instructed to hold in the top plunger a second time "until cup empties." Made by Clawson Machine Co. Pat. Apr. 3, 1894.

THREE FOR ONE

One of the first payout machines is businesslike in design and primitive in operation. If nickel safely passes "1," it must also avoid "2" as only the "3" slot trips the mechanism to automatically return three for one. Jones Novelty Co. 1892.

CARLO

Pushing the plunger drops the penny onto the pin-studded wheel. As it rocks, the wheel flings the coin onto the gold-flecked field and (maybe) into one of the four jackpot slots that is released with the next play. Clawson Machine Co. 1894.

BABY JACKS

A spring lever shoots the penny onto the field so that it spins around before its descent. A lucky hit and a slow turn of the handle drops the jackpot. The case is in colorfully painted art deco style. Field Mfg. Co. 1929.

THE INVESTOR

Respectability and dignity are implied in the name and design. However, the investment favors the operator who could adjust the bottom row of pins to control the odds of winning. Made by Charles T. Maley Co. 1893.

3 JACKS

The term "jacks" is derived from "jackpots," the pots of money showing in the windows. The irresistible sight of a heap of coins waiting to be released was the main reason for the popularity of these machines. Rock-Ola Mfg. Co. 1930.

FIVE JACKS

Nine jackpots is the largest size jacks counter machine made, with single and four jacks also manufactured. Notice most Depression-era jacks machines are penny play, unlike the nickel turn-of-the-century models. Rock-Ola Mfg. Co. 1930.

YANKEE

A side plunger flicks the nickel onto the green felt-covered field. Bouncing against the pins, the coin drops out of sight where it may land in a jackpot slot. If it does, cranking the side handle releases the coins. Caille Brothers. 1905.

ROCK-AWAY

Almost forty years after its invention by Clement Clawson, the rocking wheel device was still appealing. The player can watch the boy and girl seesaw when his penny hits the disc. Every play returns a gumball. O.D. Jennings Co. 1931.

VICTOR LITTLE SIX

Six countries share the wheel, but only America pays out—five nickels for one. This compact but fancy cast-iron one-wheeler has an etched and mirrored glass front with a bald eagle in its center to reinforce the patriotic theme. Made by Victor Novelty Works in 1907.

IMPROVED CHICAGO

In the beginning, one-wheelers did not offer a choice of bets. This early pinwheel machine has a card on its front stating the various payoff odds on the designated colors. On this model, nickels drop out from the right-hand cup, but other models would also dispense jackpot trade checks from the left. Made by D. N. Schall Co. in 1896.

BROWNIE

Not just a modification of a one-wheeler, the complex mechanism makes this counter model comparable to an upright slot. It is named for the popular sprite who appears dressed in different costumes on the lithographed tin wheel. Mills Novelty Co. first made the Brownie in 1903.

BEN HUR

Economy of space and cost are not the only advantages these countertop models have over the upright one-wheelers. A Caille Brothers Co. catalog recommends this machine for towns closed against larger machines since the Ben Hur can easily be whisked away under the counter, as it is only 21" high. 1908.

SQUARE DEAL

Striving for novelty, many manufacturers turned the wheel so that the edge faces outward. These variations are now known as "one-reelers." Here, Mills Novelty Co. allows the player to bet on a single country by moving the front indicator. True to its name, the machine provides gum with each play when the patron slides the lever. 1911.

CHECK BOY

Knight, dragon, lion, falcon, bull, eagle—up to six people can select from these symbols on the reel. When the reel stops spinning, flags pop up in the marquee to announce the winner and payout. Ten cents to two dollars in nickels fall into the cup on the right side. If the operator preferred, trade checks could be played. Mills Novelty Co. 1907.

SILVER CUP

Double the excitement for up to five players. The selected color on both dials must stop at the center arrow to pay out. The jackpot symbol is a loving cup suggesting awards, and this motif is repeated in the gleaming cast-iron front. Made by the Caille Brothers Co. in 1912.

THE DUO

The time-consuming color selection process has been eliminated. The game here is to match colors at the tangent of the two wheels in order to win. When white meets white, the visible jackpot empties. Its window is flanked by the Charles Fey & Co. logo and an Indian in the cast-aluminum front. 1910.

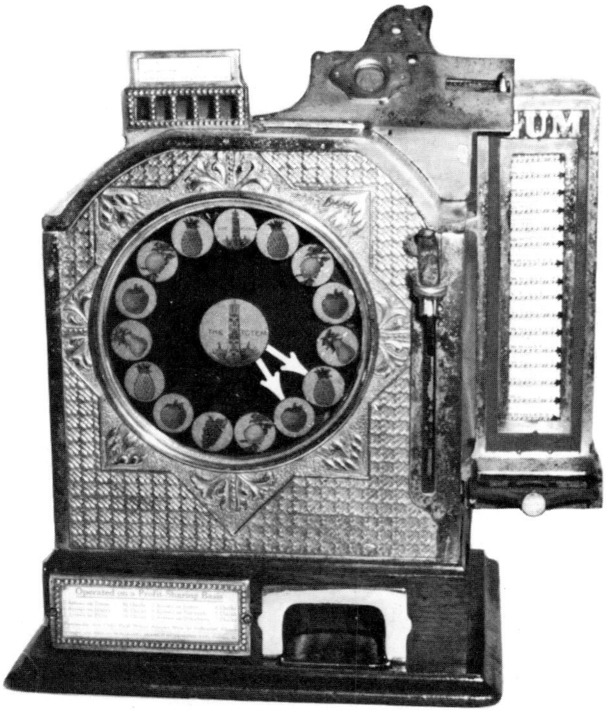

TOTEM

Two clock hands spin on the face of this ornate cast-iron and oak case. If the hands stop at matching fruit symbols, a flag in one of the little windows indicates the payout due on the next nickel. If both stop at 12 o'clock on the totem pole, the next coin gains the two dollar jackpot. The side lever must be pushed to get the gum. Watling Mfg. Co. 1911.

LITTLE CHIEF

Competition stimulated imitation, and manufacturers were not adverse to copying successful themes, styles, or mechanisms. Charles Fey's spinning arrows machine features Indian good luck symbols on its dial. Four Indian chiefs watch the player from the cast-iron front. The arrow machines' novelty was short-lived when players opted for three-reelers. 1910.

TOTEM

Three arrows whirl in two directions and stop one by one at Indian symbol decals. The wheel is flanked by magnificently detailed totem poles in the casting. The marquee lists the combinations that return the "dividend" falling due with the next nickel. This system theoretically obviates the charge of gambling, as does the gum vender in the base. Mills Novelty Co. 1911.

LIBERTY BELL

This is the one that started it all. The first of today's familiar three-reelers was designed in 1905 by Charles Fey. From the operator's standpoint, the reels offer speedier play for higher profits. The player enjoys the suspense and the mesmerizing effect of the revolving wheels. If he wins here, the player also gets free drinks.

OPERATOR'S BELL

The customer is not really gambling when he buys Liberty Fruit Gum by depositing a nickel in the slot and pulling the handle. The proper combination of symbols on the reels earns a share in the profits from the gum sales. If he also wants to chew the gum, an attendant provides it. Caille Brothers Co. 1910.

OPERATOR'S BELL

Mills Novelty Co. was the first to use today's standard fruit and bar symbols, copyrighted with this machine. Fey's basic mechanism was improved by the gooseneck coin slot and slug detector window. Most machines offer well-designed fronts, but even the sides could attract the eye. This patriotic example shows the Liberty Bell above the Mills Owl and the Statue of Liberty in New York harbor. 1910.

PACE'S RACES

And they're off!
Cheer Man-O-War or Plucky Girl as they gallop down the gleaming track in 20 seconds. The odds change after the horses leave the gate, and a numbered flag pops up at the end of the track to tell the winner. Two to thirty nickels drop into the hidden payout cup in the cabinet side. An electric motor powers a pneumatic system using a perforated paper roll to control the bellows. When Pace Mfg. Co. introduced this first successful electric console in 1934, it soon became the biggest moneymaking machine of its day.

HORSE RACE

Seeburg Corp.'s countertop horse race offers a simply designed cabinet, front knob for selecting the horse, slug detector, and a window at the edge of the turntable to show the changing random odds on each race. 1936.

SPARK PLUG

Choose a horse by positioning the front dial. Insert a nickel and push the handle to whirl the disc mounted with eight tin painted horses. The one stopping in front wins. Bally Mfg.'s early product pays a random reward of 1 to 20 coins. 1934.

RACES

Horses spin on three concentric discs, allowing for many payout combinations. Superior Confection Co. promotes its mints by use of the front vender, and offers additional play incentive with a gold award token payout. 1935.

RELIANCE

Sophisticated mechanization imitates a realistic crap game. The left window holds the first toss, relieving the player of the need to remember the point. For the same 25¢, he may continue to pull the handle to shoot dice into the right window till he craps out or wins. A $25 jackpot token is released for four consecutive 7's or 11's. Bally Mfg. Co. 1936.

DICE

This machine has all that any high-roller could ask for. With each roll of the dice under the round window, he may place additional bets on the next toss. The increased cost of playing is offset by the heightened interest in the outcome of every throw and the increased size of possible winnings. Mills Novelty Co. 1935.

LITTLE ROULETTE

Highly skilled engineering was required to produce the compact mechanism in this 19″ version of the large floor-model roulette. When the player cranks the handle, the wheel spins to propel the ball to the outer rim. As the ball slows and drops back, the mechanism brakes. If Lady Luck smiles, the winner is paid. Caille Brothers Co. 1905.

DOMINO

The dominoes spin within the pictured barroom. If the lucky barfly hits a proper combination, the painted bartender and patron raise their glasses in a toast as the machine automatically pays out. No doubt this clever variation of a three-reeler stimulated drinking as well as gambling. Distributed by the Domino Co. 1928.

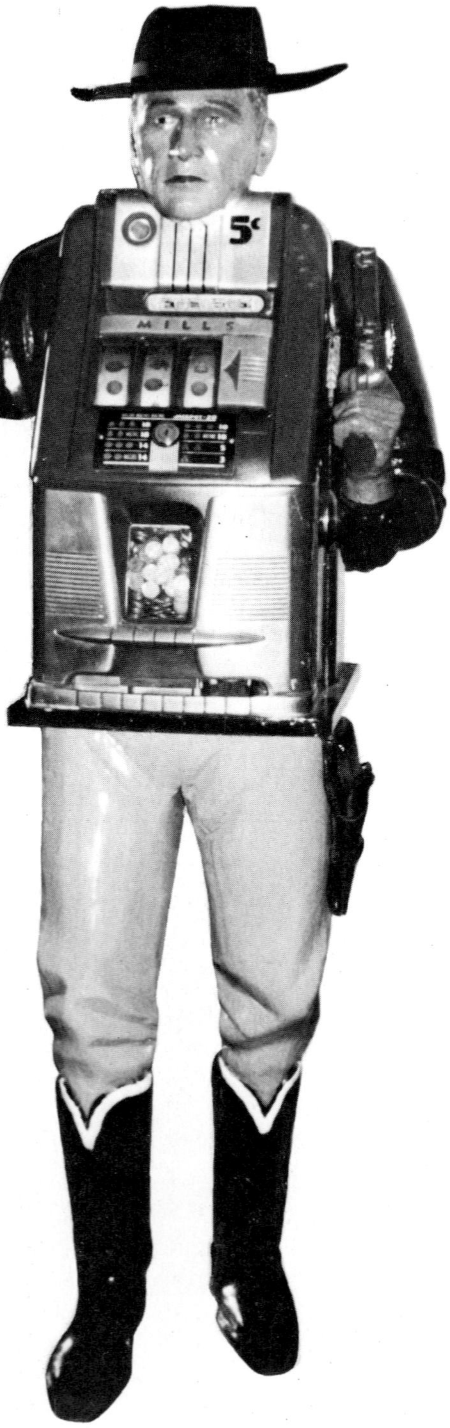

ONE-ARMED BANDIT

The one-armed bandit was brought to life by Frank Polk's skilled carvings of over 70 stock Western figures. The showdown begins when you lower the bandit's arm until his six-shooter aims squarely at your chest. Gaze into his piercing eyes or watch the reels spin in the Mills High Top, painted to match the shirt. Even losing is fun when you stand face to face with a life-size desperado. The Polk figures were produced for Nevada Novelty Co. in the early 1950s, and similar reproductions were made in 1978.

O.K. VENDER

According to Mills Novelty Co., because of the future-pay and gum-vending features, this machine was "pronounced absolutely legal by the best authorities." However, it cost an extra nickel to collect one's winnings, and the serious chewer had to move fast to remove his gum before the cabinet swallowed it back. The player also learns his fortune by checking his combination against those listed on the front. The cast-iron and oak comprising the machine's exterior are beautifully harmonious. 1912.

BABY BELL VENDER

If the player deposits a nickel to try his luck on the reels, he may also purchase Blue Seal mints by turning the side knob. But if he plays with a token, all he can hope for is more tokens. Made by Watling Mfg. Co. in 1927.

TREASURY

This famous example of Watling styling has heaps of coins in the casting, which flank the real thing in twin jackpot windows. Variations have a vender or gold-award token payout. In any form, it's a popular machine today. 1936.

BALLGUM FRONT

One-cent play and colorful gumballs attract the casual player. He has little to lose and as much as 40¢ to gain if he hits the jackpot. In any case, he may always learn his fortune printed on the reels. Watling Mfg. Co. 1932.

LINCOLN DE LUX

Gambling devices often hid behind patriotic sentiment, so the esteemed president's profile here provides respectability. Other features to note are the Watling logo medallion on top, and magnifying-glass slug detector. 1926.

ROL-A-TOP

The famous Watling horn of plenty, spilling its fruits of prosperity, curves around the eight-coin round escalator. Some versions have a mint vender or gold award, and the cornucopia holds cherries or coins. 1935.

BABY BELL JACKPOT

Watling Mfg. made an entire line of Baby Bell slots, which are still respected for their sturdy and dependable mechanisms. An unusual feature on this model is the lure of two plays for five cents. Note also the typical "gooseneck" coin slot. 1932.

VICTORY

The center handle saves counter space, but it's awkward to operate, and no other machines repeat this feature. This ornate slot also has future pay and a mint vender to appease local guardians of morality. Caille Brothers Co. 1920.

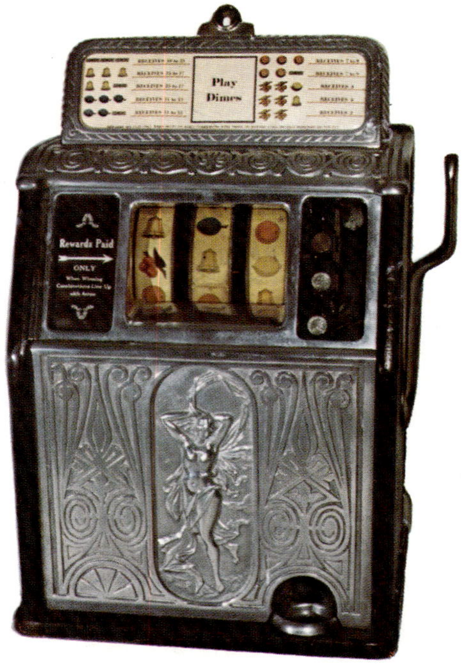

SUPERIOR

This lovely case houses Caille's unusual slug detector. Dimes dropped in the slot fall to the bottom, are picked up by the escalator, and carried upward past the window, entertaining the player and protecting the operator. 1926.

GRAND SLAM

Caille Brothers added a fourth reel to its exotic Sphinx cabinet. The more familiar reels give the usual payout odds as indicated on top. The extra reel is needed only to win the jackpot. 1932.

1927 OPERATOR BELL

This is the first of a long line of successful Pace gambling machines. The design broke no new ground, but is well made and attractive with its rotating slug detector, ornate payout card, and prominent jackpot.

COMET

"The Comet blazes the way to greater machine profits." The operator could preset the amount of coins that would drop into his cashbox or into the twin jackpots, which alternately pay out to keep the machine in action. Pace Mfg. Co. 1932.

COMET VENDER

This version of the classic Pace Comet includes a mint vender for sensitive locations. But few players who tasted the poor-quality candies once would bother to remove another pack with their next play. 1932.

SILENT GOOSENECK

The "Lion Front" slot machine's bold design attracted much attention when the machine was introduced in 1931. With its brightly colored front and a backup reserve jackpot, it was another Mills Novelty Co. success.

O.K. MINT VENDER

"It sets on your counter stimulating sales and making big profits, without any attention or care," claims a Mills Novelty Co. ad in 1929. It also vends mints automatically and "at occasional intervals" pays a reward.

BONUS BELL

The first reel contains the usual fruit symbols, and the letters B-O-N-U-S. Hitting the letters in order earns an extra payout. The more previously hit letters showing, the more tempting it is to play. Mills Novelty Co. 1937.

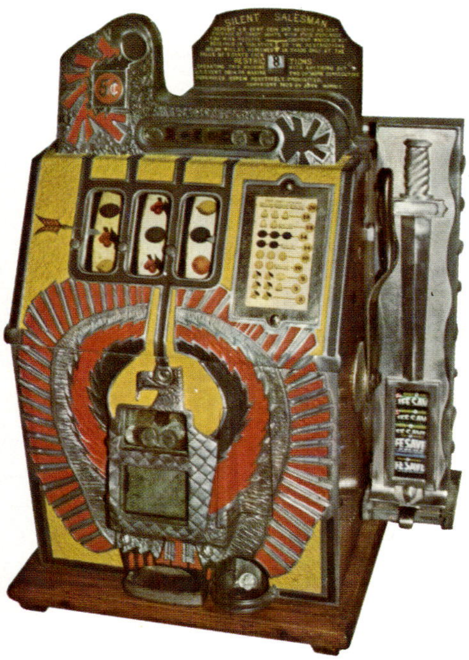

SILENT JACKPOT BELL

"The War Eagle" is emblazoned in bright enamel across the cabinet front, and inside is the new smoother "silent" mechanism. This first machine to unify design and theme introduced the horizontal escalator. Mills Novelty Co. 1931.

MYSTERY BELL

The medieval knight, with his sword and shield, captured the player's interest for many years. Known in the trade as the "Blue Front," this slot machine was introduced by the Mills Novelty Co. in 1933.

BASEBALL O.K.VENDER

Baseball symbols on the reels camouflage a less innocent activity. Scoring dials and the scene on the front are also part of the elaborate coverup. In case no one was fooled, the mint vender gave additional protection. Mills Novelty. 1929.

VICTORIA

Peacocks strut on either side of the stylized tail that surrounds the jackpot window. The design of this cast aluminum and oak machine has a harmony not always found in three-reelers. Made by O.D. Jennings Co. in 1932.

LITTLE DUKE

Three flat concentric discs make this art deco classic a unique variation of the standard three-reeler. The center column of symbols appearing in the pie-wedge-shaped window determines the payoff. O.D. Jennings Co. 1932.

TO-DAY

The Dutch boys invite the patron to try their tempting mints, while the visible jackpot entices him to try for other rewards as well. A rotating drum on top reports the next payout, if today is the player's lucky one. O.D. Jennings. 1929.

A.C. MULTI-BELL

After the sale of the Caille Brothers Co., Adolph Caille formed the A.C. Novelty Co. and produced this slot. But the multiplay and 7 separate jackpots were complex and unreliable, so the machine (and company) were discontinued. 1936.

CHIEF

Silver-dollar play provides richer happy-hunting grounds when the jackpot is hit. First used with this machine, the Indian became a popular O.D. Jennings mascot for many years. 1936.

COUNTER O.K. VENDER

To operate this cast-iron and oak paragon, the player drops a nickel in the head, squeezes the top lever, and pulls the handle. Winning combinations tell his financial fortune, but he is entitled to gum just for trying. Mills Novelty Co. 1923.

VEST POCKET

The smallest full payout slot is 8″ high. When authorities drop in, the flap covers the mini-reels. Another disguise is provided by eliminating the handle—play is activated by simply pulling out the coin slide. Mills Novelty Co. 1938.

THE BABY GRAND

C & F Mfg. made the first three-reel payout machine in a miniature cabinet, and even included a jackpot. Only 12″ high, it occupies less counter space and can easily be tucked away at inopportune moments. 1931.

MIDGET

This space-saving slot machine employs only two standard size reels and stands 17″ high. Burtmeier Co. used a longer spin to maintain interest. 1935.

Q.T.

Because of its compact size, Mills Novelty claimed in 1934 that operators would say "O.K. to the Q.T." The three-reeler even has a twin jackpot and bent coin release. This model with speckled and shaded paint was made in 1948.

COLUMBIA DELUXE

The gold anodized aluminum makes this little three-reeler shine. Coin denominations could be changed by switching a few mechanical parts. In spite of this, the mechanism was generally unreliable on location. Groetchen Tool Co. 1936.

BALLY DOUBLE BELL

Bally Mfg. Co., today's largest producer of slot machines, began its slot manufacturing with this machine. Two slots in one, separate systems allow the hotshot gambler to play a nickel and quarter at once to try for two jackpots. 1937.

Trade Stimulators

Trade stimulators are gambling devices which do not automatically pay out cash awards. Instead, the payout is a token good for cash or merchandise, or merchandise itself or cash delivered by an attendant. Trade stimulators can be regarded in any of three ways: 1. They are exactly what their name implies — innocent machines, which in the spirit of competition and free enterprise, are designed to entice the customer to patronize the store. 2. They are wicked deceptions made to camouflage their real purpose — to encourage the nefarious and illegal activity of gambling. 3. A combination of 1 and 2.

In a 1911 Caille Brothers Company catalog, the success of the trade stimulator is summed up as follows:

> "These machines appeal to the known spirit of speculation, so strong in every man. They interest and gratify that element in him at small cost to himself, and they give him a fair show and "a good run for his money" against slight odds.
>
> "The merchant who is anxious to increase his business can do so by means of these machines. They are made expressly for the purpose of stimulating trade in a manner equally satisfying to merchant and customer."

Merchants also liked them because they were simply designed and did not easily "get out of order," and because they were inexpensive to buy (some sold for as low as $2.50). Sometimes they were even offered free by product manufacturers if the storekeeper purchased a minimum amount of their product. Usually located near the cash register, trade stimulators proved to be an additional source of income.

Some of the machines had mechanisms designed to automatically reward not coins but trade tokens good for a specified amount of merchandise. Others offered "fortunes" which were cleverly worded to indicate the amount payable by an attendant whenever a winning combination came up, (pp. 43, 50, 51). "Honest" or "fair" stimulators pretended to be venders and guaranteed to pay merchandise, usually cigars, for every nickel played. Rewards would be two or more cigars for the same nickel, if the player was lucky. (The question as to the quality of a "nickel cigar," however, was left to the discretion of the merchant.) Often the machines came with assorted pay cards giving the merchant the option of offering liberal or moderate rewards in either merchandise or cash for either a penny or nickel.

The trade checks themselves were stamped pieces of metal, frequently made in distinctive and attractive shapes to ensure that they could not be passed off as real coins. Merchants could order checks with their own names; sometimes the name of the machine and/or manufacturer was supplied. Other trade checks simply indicated a coin denomination for which the check could be cashed in. Some machines were designed to accept checks in addition to coins, while others were made to reject them.

Trade stimulators were designed in a variety of unusual shapes, styles, and themes to intrigue the consumer. In all cases, the satisfaction derived from receiving the merchandise seemed to be no less than when coins were paid out. With trade stimulators, it was apparently not how much you won or lost, but whether or not you won at all.

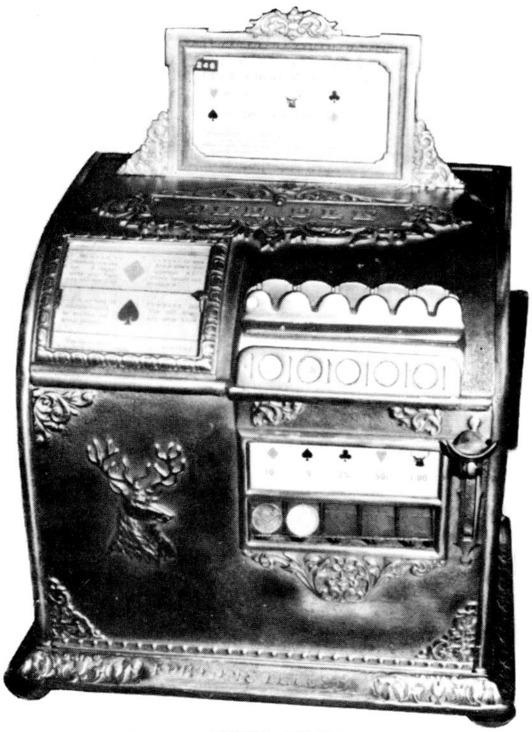

THE ELK

Is there a cigar in your future? Ask this gleaming copper-flashed cast-iron fortune teller by inserting a nickel in any of the slots and pressing the handle to spin the reel. When it stops, you may read your fate. If it also shows your spot, a bell rings as a token is dispensed. Turn this in for your smoke. Paupa & Hochreim. Case design pat. June 21, 1904.

LITTLE SCARAB

Busy scarab beetles, painted red and green, dance about in the shiny casting of this countertop roulette. Place your bet and pull the lever to spin the wheel. If the metal ball lands on your choice, a trade check is automatically ejected into the little cup on the right side. Mills Novelty Co. 1910.

BASEBALL

"As popular with the public as the game itself." Caille Brothers took advantage of the passion for the national pastime and converted its Tiger model by changing the pay-card and front casting. Colored playing fields designate the number of base hits (and the value of the token) but the 30¢ top payout is still represented by a tiger. 1908.

UMPIRE

Will you hit a homer or strike out? Mills Novelty Co. included strike, foul, and out symbols on the reel so that even if all five colors were played, there still might not be a pay-out every time. The catcher in the copper-plated front casting appears to be reaching for the pennies in the window, which shows the bets placed. 1905.

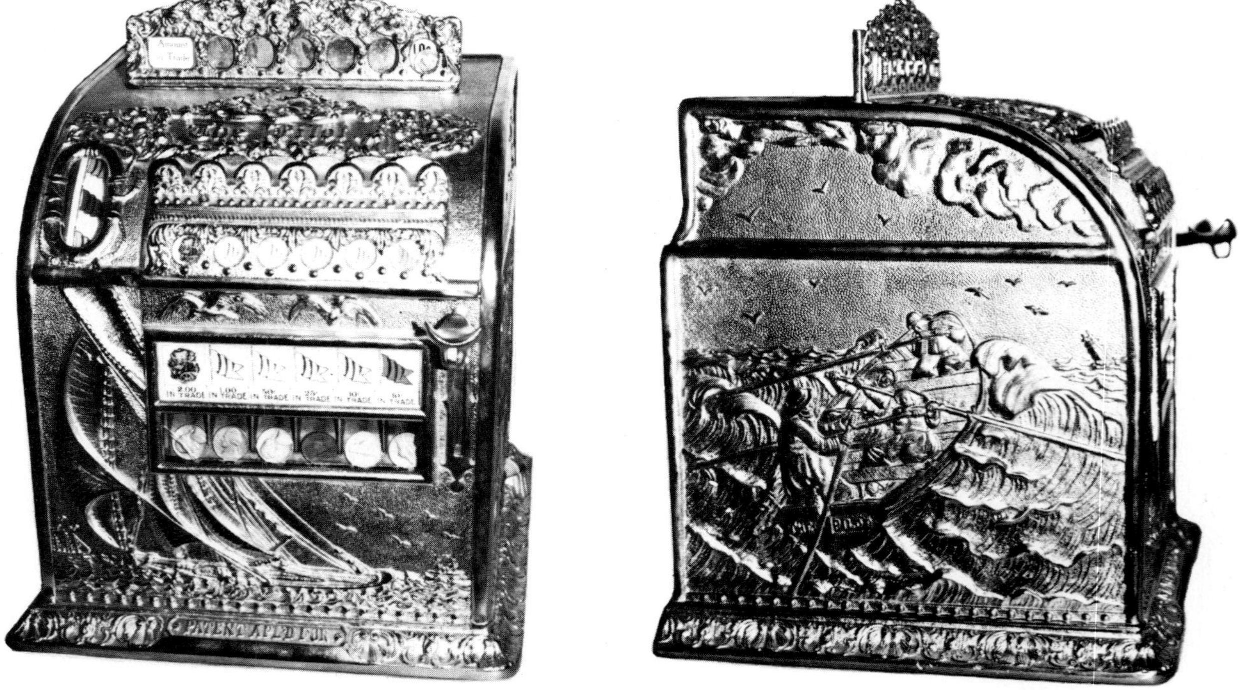

THE PILOT

Steer your way to the jackpot. Drop 5¢ in any or all coin slots, press the handle, and watch the reel spin. If you win, a flag or the jackpot symbol of a pilot pops up in its window on the marquee, a bell rings, and a trade token (worth 10¢ to $1.00) drops from the side. Early cast-iron trade stimulators were often richly detailed. Here, the elaborate scene on the side depicts two oarsmen on a raging sea. Mills Novelty's design and craftsmanship make this an object of art. 1906.

THE ELK

In 1905, big game hunting was a fashionable attraction for the player. And if one company felt another had a winner, it readily copied. Mills Novelty's Elk closely resembles that of Paupa & Hochreim – in motif, its offer of free cigars or drinks, and the fortune-telling reel. However, the well-known owl trademark and the elaborately detailed sides help distinguish theirs from the competition.

CRAP SHOOTER

Mills Novelty thriftily converted their Pilot case to create a dice game. The player selects his bet from the six on the card by inserting a nickel, dime, or quarter in the right slot. Pressing the handle rolls the dice. 1906.

AUTOMATIC DICE

This mechanism is fascinating. A clockwork motor closes the tops over the dice, turns the cups one revolution, and opens the lids to show the toss. The top paycard lists the spot totals meriting reward. Clawson Machine Co. Pat. Mar. 20, 1893.

ON THE LEVEL

No need to worry about loaded dice. This cast-iron craps game can accommodate up to six players. The attendant checks a window in the back to see the coins played and pays rewards to the winner. Charles Fey & Co. 1907.

CRAP SHOOTER'S DELIGHT

Dice, the country's most popular gambling game, was one of the earliest trade stimulator themes. A player of this antique-copper machine reads the sides of the dice, not the tops, to determine if he has won any cigars. F.A. Ruff Co. 1905.

I WILL

Two voluptuous nudes grace the front of this copper-plated dice game. Each glass, embossed with the Mills owl, holds five dice. A twist of the handle bounces all the dice, but winnings are determined by the total spots in the glass bet on. 1904.

WINNER DICE

Everybody wins, even if it's only a package of "delicious quality mints." Certain totals return additional rewards. Caille Brothers copied this glass-domed cast-iron game in 1905 from an 1892 U.S. Novelty model.

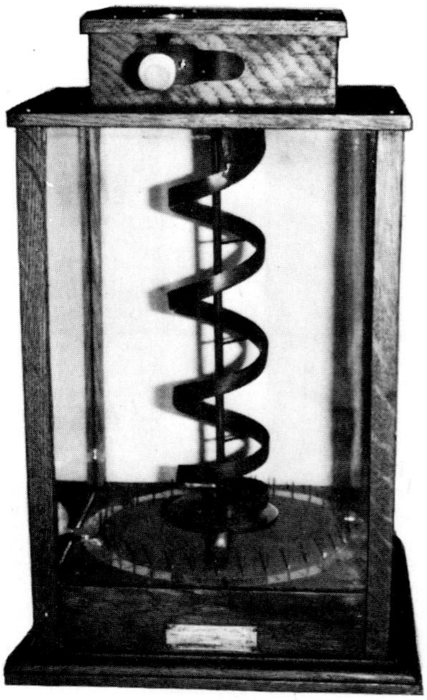

SPIRAL

Watch a nickel gyrate out of sight. Pushing the rod turns the corkscrew chute as the nickel travels down. The flag at the foot of the spiral spins. The number on the base it stops at decides the reward. 20th Century Novelty. Pat. Mar. 13, 1906.

WIZARD CLOCK

A real fooler in function and appearance, this mantel clock only bongs for money, not on the hour, paying up to three trade tokens for each coin dropped in top. The wood is painted to imitate ebony and marble. Wm. M. White Co. Pat. Aug. 14, 1906.

LIVELY CIGAR SELLER

Heads or tails? Flip against the proprietor for 5¢ cigars. Drop five pennies (or a nickel) in the coin slots at the top and press the lever. After landing via the bent wires, all heads wins extra cigars. Clawson Machine Co. 1895.

THE WESTON

A delicate clockwork mechanism tilts the two trays back when a nickel is deposited. When the trays straighten, metal balls roll into the numbered channels. The player compares his total with those on the paycard. Weston Slot Machine Co. 1893.

AUTOMATIC PHRENOLOGIST

Instead of reading the bumps on your skull, this mechanical phrenologist reads the "wheels in your head." The numbers at the indicators reveal your character traits, some of which could win cigars. Norman Lichty Mfg. Co. Pat. Jan. 2, 1894.

ECLIPSE

Win up to ten dime cigars. This early gambling device has a clockwork motor to spin the two arrows. When they stop, the total of the numbers they point at tells if and how many cigars you have won. The Anthony Co. 1893.

KLONDIKE

The pretty handcarved oak cabinet displays a multicolored pinwheel. The arrow spins when the coin is inserted into a color slot. If the prospector strikes gold, he is rewarded with merchandise. Charles Fey & Co. 1897.

WHEEL OF FORTUNE

This early example of a dial trade stimulator is based on the game of roulette. A ring of numbers and directions appears on the glass front. A color wheel revolves behind. Rewards are paid in 5¢ cigars. Griswold Mfg. Co. 1893.

WASP

This multicolored marbleized cast-iron trade stimulator spins on its turreted base to face any player. A nickel in the marquee and a push of the handle spins the wheel with its five pompadoured beauties. Caille Brothers Co. 1905.

MIDGET

"Midget" refers to its diminutive 9," and "Three Musketeers" to the game with three balls spinning on the wheel under the dome. If each ball lands on the same color, an attendant pays the reward in trade. Charles Fey & Co. 1928.

LITTLE MONTE CARLO

Instead of traveling to the famous Monaco casino, the less-than-wealthy playboy may enjoy this "perfect miniature roulette" for a nickel. More conservative gamblers could try the penny version. Made by Mills Novelty Co. in 1905.

PURITAN BELL

This sturdy little machine resembles a cash register in shape and a three-reeler in its standard slot machine symbols. Winning hits were paid in cash or trade by the clerk. Many companies made variations; this one's by Mills Novelty. 1922.

THE BICYCLE

The bicycling craze of the 1890's inspired unusual trade stimulators. The charm of a 7" high bike in a handmade oak case and the suggestion of healthy sport made this classic machine welcome even in sensitive locations. But the only exercise the player gets is when he drops in his nickel and turns the lever. A bell rings and the tires spin. Indicators point to a number on the rim of each wheel. The player then compares his total with those on the paycard to see if he won. Waddell Wooden Ware Works. 1893.

FAIREST WHEEL

Fair play is assured. The player is guaranteed one cigar for his nickel, and has a chance to win two. The weight of the coin hitting the edge causes the wheel to spin on its delicately balanced bearings. Fairest Wheel Co. Pat. May 7, 1895.

BICYCLE WHEEL

"The best 5¢ cigar in the house for each nickel dropped in the slot" is a promise, and two or five are a possibility. The name is etched on the glass, and delicate basket-weave fronts the arched oak case. Waddell Wooden Ware Works. 1895.

STAR ADVERTISER

The weight of the dropped coin spins the needle. The "best" cigars were not always worth five cents, but the chance of getting more than one kept the nickels dropping. Drobisch Brothers & Co. Pat. Feb. 17, 1897.

UPRIGHT PERFECTION

Nothing less than two pair pays off, but the player willingly risks a penny ante for a nickel cigar. If he hits a royal flush, he wins 25 cigars. Made of nickel-trimmed polished oak by Mills Novelty Co. 1901.

SUPERIOR

Superior in appearance and efficiency, this handsome poker game appeals to the busier proprietor. A winning player pulls a handle to receive a printed ticket of his hand, which he turns in at the counter for the reward. Mills Novelty Co. 1905.

COMMERCIAL

Each hand is dealt from a 50-card deck that revolves under the curved glass each time a nickel is inserted. Hand rank determines the pot of 1 to 100 cigars. Dragon heads repeat the dragon motif cast in the sides. Mills Novelty Co. 1903.

GLOBE

The wealth of elaborate scrollwork on the front of Caille's unusual drum-shaped machine flows over the back and sides as well. More visual interest is supplied by each reel spinning in an opposite direction. 1902.

JOCKEY

Three coinslots and three visible poker hands allow for triple play and possibly three winners. The five-card symbol in the front casting is repeated on the art nouveau marquee framing the rules card. F.W. Mills Mfg. Co. 1905.

DRAW POKER

Card sense counts. After the first deal, the player holds his best cards by pressing the corresponding buttons on top. In hopes of drawing a winning hand, he pushes the lever again. Made by Mills Novelty Co. around 1900.

HIGH STAKES

The three horse segments must match in color. The second reel tells if the horse wins, places, or shows, and the first reel tells the amount won. The bright orange wrinkle-finished case also dispenses a gumball each play. Groetchen Tool. 1935.

OFFICIAL SWEEPSTAKES

The player selects one of the brightly colored entries and squeezes the knob to start the race. The turntable spins, kicking a ball around its numbered edge to determine the odds to be paid on a winner. Rock-Ola Mfg. Co. 1935.

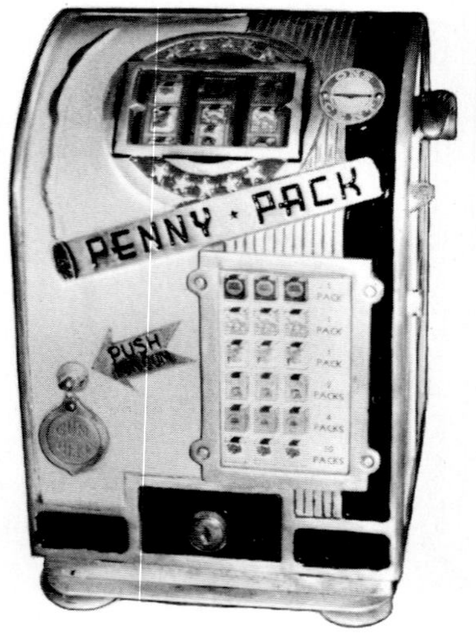

PENNY-PACK

By 1934, most smokers had switched from cigars to cigarettes. Therefore, the lure of a pack (or ten) for a penny, by lining up the same brand on the reels had great appeal. This typical art-deco style cabinet was made by Daval Mfg. Co.

PURITAN BABY VENDOR

"A family of four will bring you happiness." Payout amounts are cleverly disguised by including numbers in the fortunes. A clerk checks the side window to see the coin played, and pays the proper multiple. Daval Mfg. Co. 1931.

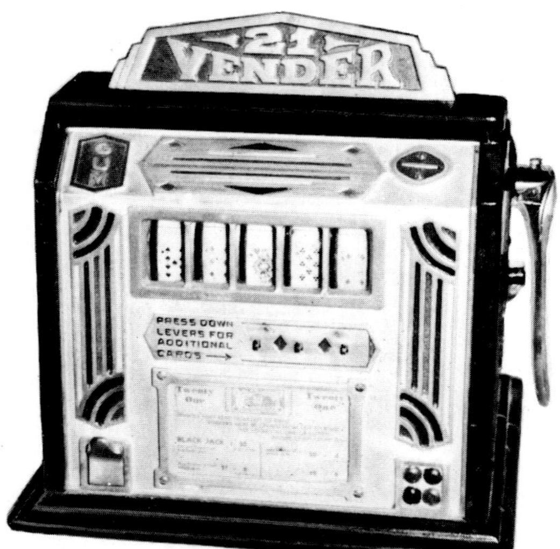

"21" VENDER

Should you stand or be hit? Keep drawing till you get 19 or better. If you do, point amounts are indicated on the front paycard. At any rate, you are entitled to a gumball with each play. Groetchen Tool Co. 1930.

WHIRLWIND

See what the winds of chance will bring, besides the ball of gum. The jackpot dumps automatically, but other combinations were paid at the counter. Made by Pierce Tool & Mfg. Co. in 1933.

ONE CENT MERCHANT

·Support democracy by buying anti-trust gum, and help your own finances by hitting a winning combination. Caille Brothers used unusual lithographed fruit symbols on the reels and put their logo in the scrollwork. 1905.

BABY VENDER

The trick of payout amounts concealed in the fortunes was effective enough to be copied in trade stimulators of different designs and materials. This popular style is made of oak with a cast-aluminum front. Southern Doll Mfg. Co. 1925.

TRY YOUR LUCK

Everyone wins. The operator received this machine free with a sufficient purchase of gum. The manufacturer, William Wrigley Jr. & Co., sold more gum. And the player had a chance to get more than his penny's worth. Pat. June 22, 1897.

THE FAVORITE

You'll need exceptional skill to get more than a gumball. Tap the trigger to shoot your penny into the high-point bull's-eyes or past the nails into lower scoring pockets at the bottom. O.D. Jennings. 1928.

PREMIUM

This cast-iron penny drop outwits the gambling laws by awarding a stick of gum for every coin that drops into a "G" slot. If the player overcomes the probability laws, he may also receive extra merchandise. Mills Novelty Co. 1911.

TARGET PRACTICE

Mills Novelty Co.'s version of the penny flip guarantees the player a piece of gum for trying. If he's skillful enough to hit a numbered slot or target, the management would pay in merchandise or cash. 1926.

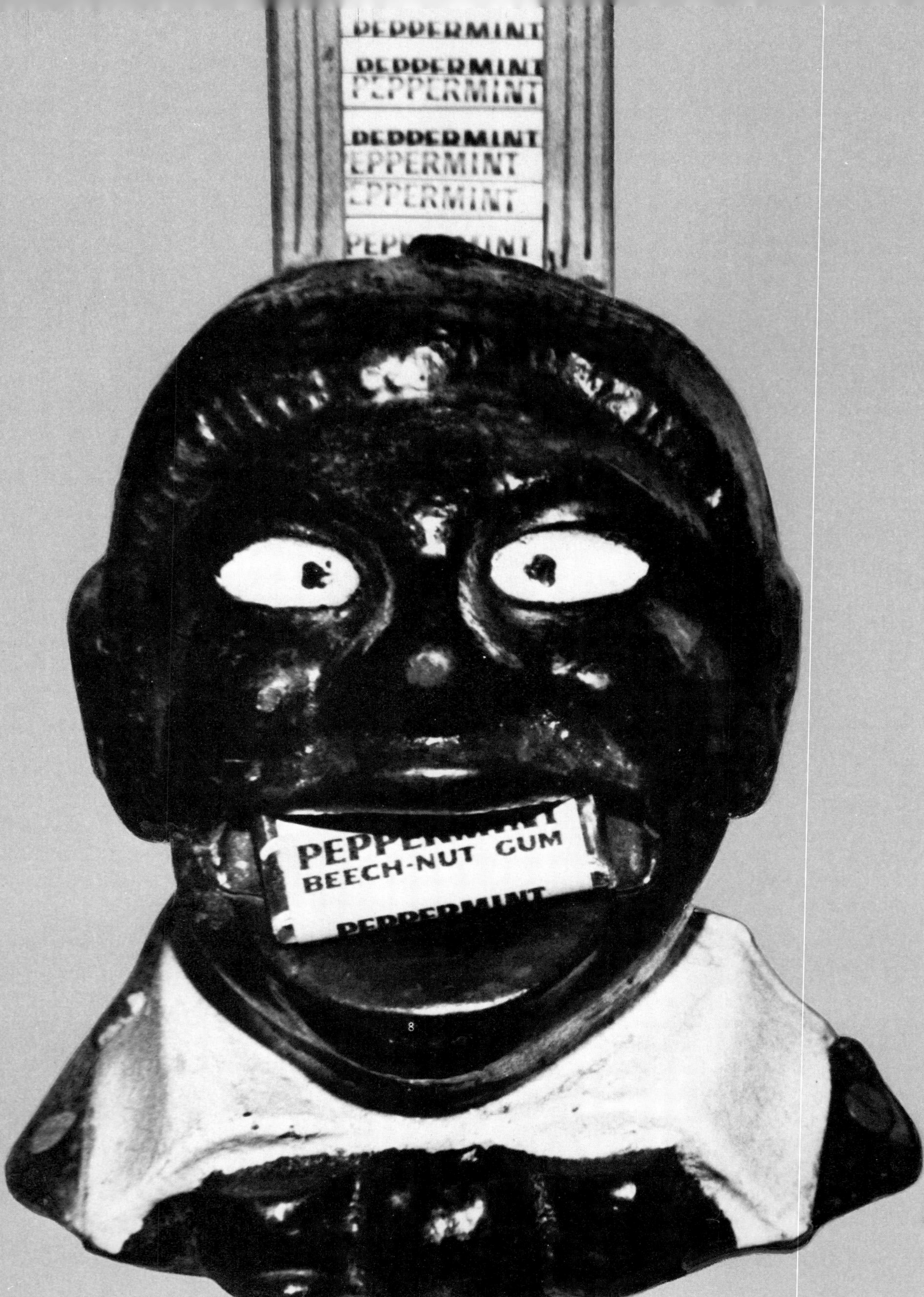

VENDING

By 1890, vending machines were already a permanent fixture on the American scene. In just a few short years, the public had learned to appreciate the convenience and to delight in the novelty of making a purchase from a mechanical salesman.

Although the first coin-operated machine had appeared here in 1839, it was not until 1884 that the first American patent on a vending machine was granted. Called an "Automatic Liquid-Drawing Device," W. H. Fruen's invention was, for some reason, made in the shape of a three-story townhouse. The cup was kept on the stoop, under the front door.

309,219. AUTOMATIC LIQUID-DRAWING DEVICE. WILLIAM H. FRUEN, Minneapolis, Minn. Filed Jan. 28, 1884. (No model.)

To Thomas Adams, however, goes the honor of first introducing the vending machine to the American public on a large scale. In 1888, he installed machines on the elevated subway platforms in New York City to sell his Tutti-Frutti gum (p. 62). These machines, placed where large numbers of people were guaranteed to be present, attracted the pennies of the impulse buyer so successfully that venders for an extraordinary variety of products and services were soon established as permanent features on the commercial scene. In 1897, Pulver Manufacturing Company added additional novelty by selling their gum in venders equipped with animated figures. Mills Novelty Company brought out their first peanut vender in 1902 to merchandise their own brand of Spanish peanuts. Also in 1902, Horn & Hardart opened the first automatic restaurant in America in Philadelphia, which even sold beer and liquor from a five-cent slot. (They were so successful that a branch was opened in New York City the following year.) And in 1909, the Northwestern Corporation (still in business) was formed to market match venders. Stamp venders received early governmental sanction, machines to replace physicians sold various medicinal aids, cigar venders guaranteed a sanitary product, and delicate glass-topped machines vended gold-filled collar buttons from the counter.

No single kind of merchandise is so closely associated with coin-operated vending as chewing gum (pp. 57–63). An American product even before the arrival of the Europeans, chewing gum has now become a worldwide commodity. In 1871, Thomas Adams introduced the first commercial gum product, called "Adams' New York Chewing Gum," wrapped in colored tissue paper and formed into small tabs. Other companies soon jumped in with their own variations on composition, flavor, and shape. The names Colgan, Fleer, Pulver, Zeno, and Wrigley became synonymous with gum chewing and gum venders. Gumballs, unwrapped and candy-coated (and cheaper to produce), came rolling out in 1907. The flavor and activity of chewing were incentive enough to the buyer, but some companies claimed that "Pepsin" (a newly discovered enzyme) gum was also an aid to digestion.

Peanuts were another product owing much of its popularity to the convenient and profitable vender (p. 64). In 1902, Herbert Mills cornered the market on the world's supply of peanuts to fill the containers of his peanut-vending machines. Although his strategy did not last long, he gained the name of "Peanut King."

Confections, principally the ever popular chocolate, were sold in small packets in vending machines as samplers for the company's product, for possible later purchase in larger quantities (p. 63). Subway platforms usually included chocolate machines in addition to peanut and gum venders. Some machines offered both chocolate and gum.

Matches, before coin-operated venders, were frequently

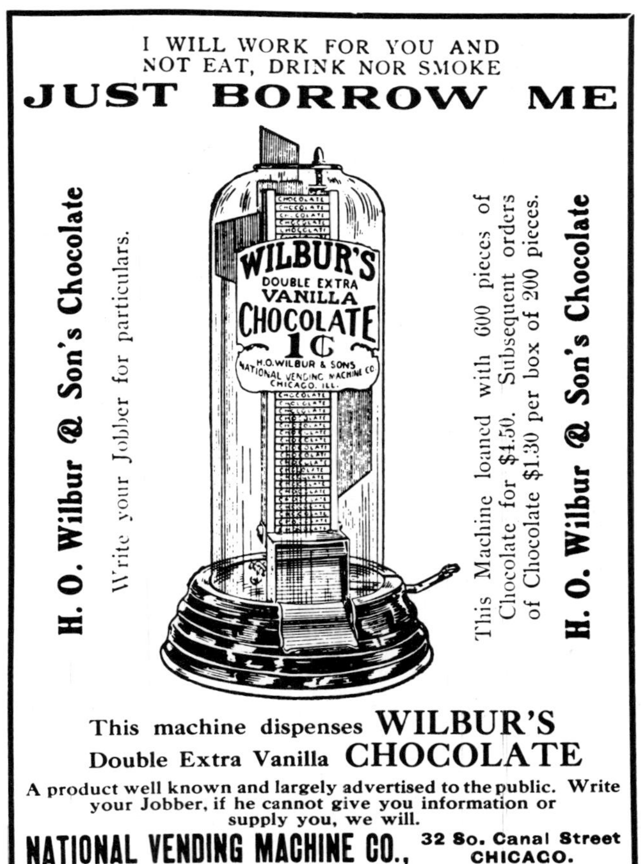

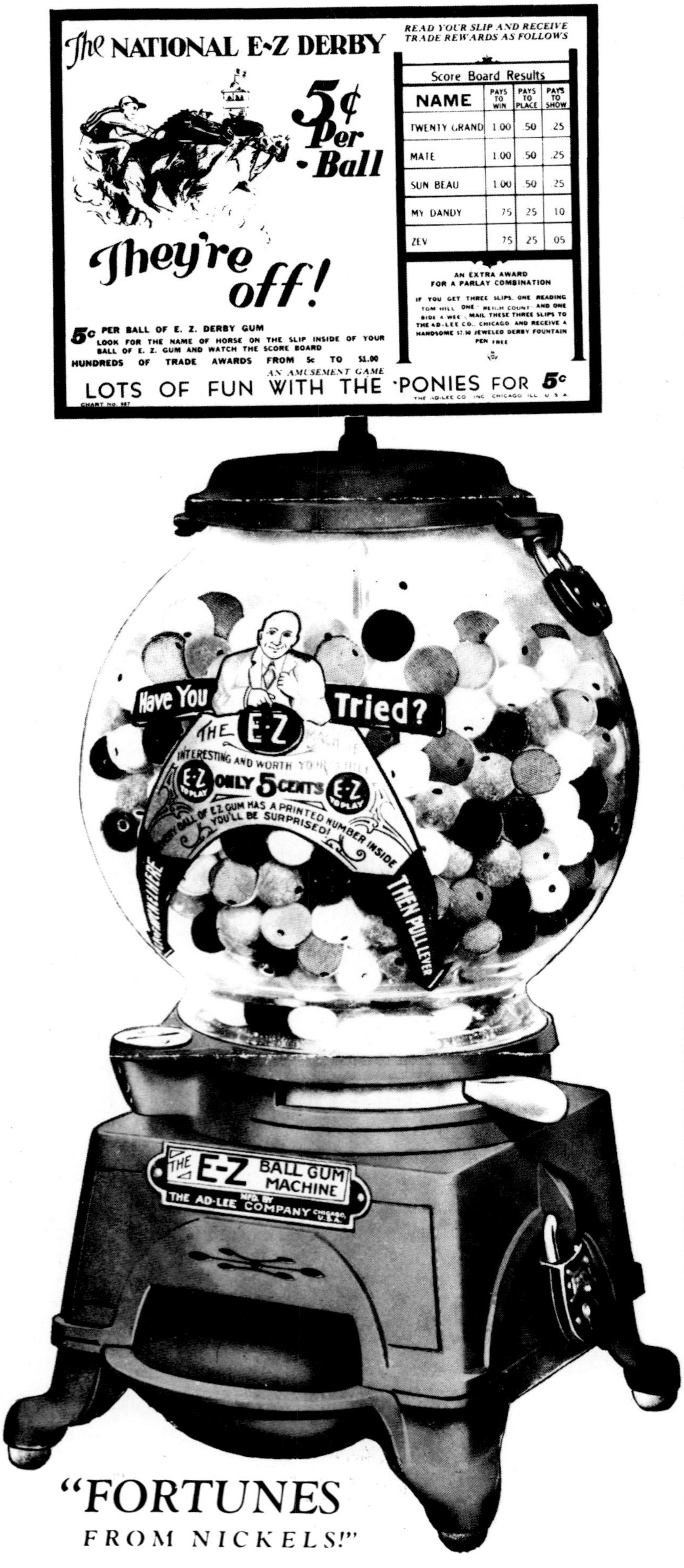

"FORTUNES FROM NICKELS!"

The National E-Z Derby 5c Ball Gum Machine

The Latest and Hottest Nickel Getter of them All

THE new ideas injected into the E-Z NATIONAL DERBY BALL GUM MACHINE have made it a push-over from the start. More prize awards than ever, besides added attractive features, yet showing THAT BEAUTIFUL PROFIT OF **$28.00** NET FOR STOREKEEPER each time the machine empties.

READ the inside pages showing how nearly **500** awards are given out to the players from each set or globe full of gum plus the ADDED PARLAY PRIZE FEATURES.

EVERY BALL OF E-Z DERBY GUM has a hole thru the center with the name of a race horse printed on the slip inside the hole. A REAL THRILL GAME that everyone will want to play.

Profits and Results for the Storekeeper on Each Set of Derby Gum Sold

1200 Balls Derby Gum sold at 5c each totals		**$60.00**
Dealer gives out candy, cigars, etc., that cost approximately	**$20.00**	32.00
1 set of Derby Gum cost him	12.00	
STOREKEEPER'S CLEAR PROFIT ON EVERY SET		**$28.00**
The E-Z Machine complete with a set of the E-Z Drilled Derby Gum cost storekeeper only		**$30.00**
Extra Sets Derby Gum, per set, only		**$12.00**

The AD-LEE CO., Inc.

The Best Vending Machines.

REQUIRE NO CLERK'S ATTENTION.

DO NOT GET OUT OF ORDER.

THE PENCIL MACHINE

IS an ornament to the counter. Requires no attention and will sell five times more PENCILS than would be sold from the show case. The PENCILS are the very best that can be made and when once used, are preferred to all others. The enormous quantity that we sell enables us to supply them at the lowest possible price.

THE CANDY MACHINE

THIS old fashioned stick candy attracts at once both young and old, and owners of the machine are amazed at the business it does. Why not add this little department in your store; it is an ornament on the counter and costs nothing for clerk hire. A grown person will drop a penny in a machine when they would hesitate to buy a cent's worth over the counter. This is one of the reasons why the machine does so much business.

FOR TERMS AND PRICES SEE OUR LETTER ENCLOSED HEREWITH

Clawson Machine Co., - - - Flagtown, N. J.

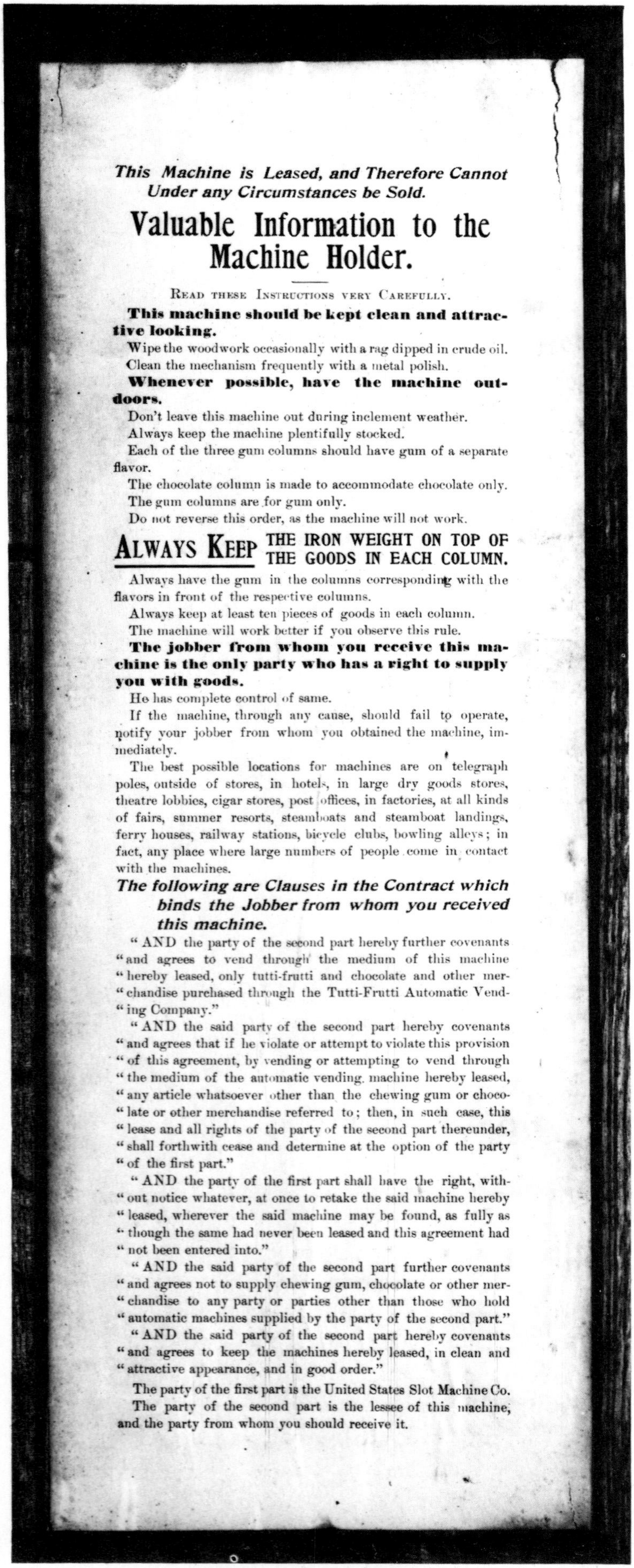

This Machine is Leased, and Therefore Cannot Under any Circumstances be Sold.

Valuable Information to the Machine Holder.

READ THESE INSTRUCTIONS VERY CAREFULLY.

This machine should be kept clean and attractive looking.

Wipe the woodwork occasionally with a rag dipped in crude oil.

Clean the mechanism frequently with a metal polish.

Whenever possible, have the machine outdoors.

Don't leave this machine out during inclement weather.

Always keep the machine plentifully stocked.

Each of the three gum columns should have gum of a separate flavor.

The chocolate column is made to accommodate chocolate only.

The gum columns are for gum only.

Do not reverse this order, as the machine will not work.

ALWAYS KEEP THE IRON WEIGHT ON TOP OF THE GOODS IN EACH COLUMN.

Always have the gum in the columns corresponding with the flavors in front of the respective columns.

Always keep at least ten pieces of goods in each column.

The machine will work better if you observe this rule.

The jobber from whom you receive this machine is the only party who has a right to supply you with goods.

He has complete control of same.

If the machine, through any cause, should fail to operate, notify your jobber from whom you obtained the machine, immediately.

The best possible locations for machines are on telegraph poles, outside of stores, in hotels, in large dry goods stores, theatre lobbies, cigar stores, post offices, in factories, at all kinds of fairs, summer resorts, steamboats and steamboat landings, ferry houses, railway stations, bicycle clubs, bowling alleys; in fact, any place where large numbers of people come in contact with the machines.

The following are Clauses in the Contract which binds the Jobber from whom you received this machine.

"AND the party of the second part hereby further covenants "and agrees to vend through the medium of this machine "hereby leased, only tutti-frutti and chocolate and other mer- "chandise purchased through the Tutti-Frutti Automatic Vend- "ing Company."

"AND the said party of the second part hereby covenants "and agrees that if he violate or attempt to violate this provision "of this agreement, by vending or attempting to vend through "the medium of the automatic vending machine hereby leased, "any article whatsoever other than the chewing gum or choco- "late or other merchandise referred to; then, in such case, this "lease and all rights of the party of the second part thereunder, "shall forthwith cease and determine at the option of the party "of the first part."

"AND the party of the first part shall have the right, with- "out notice whatever, at once to retake the said machine hereby "leased, wherever the said machine may be found, as fully as "though the same had never been leased and this agreement had "not been entered into."

"AND the said party of the second part further covenants "and agrees not to supply chewing gum, chocolate or other mer- "chandise to any party or parties other than those who hold "automatic machines supplied by the party of the second part."

"AND the said party of the second part hereby covenants "and agrees to keep the machines hereby leased, in clean and "attractive appearance, and in good order."

The party of the first part is the United States Slot Machine Co.

The party of the second part is the lessee of this machine, and the party from whom you should receive it.

Instructions found inside the Adams' Pepsin Gum machine (p. 62).

given away as an accommodation to smokers. No one could sell a single match, of course, so boxes of matches were packaged to sell for a penny. Automatic venders simply increased the convenience of purchase and probably the number of sales (p. 69). Lighter fluid was also sold for those who owned lighting devices (p. 65).

Cigars were one of the earliest products to be vended from coin-controlled machines and were staples for as long as cigars remained the most popular "smoke" (p. 68). Styles of machines and techniques for delivery were unusually clever and imaginative. By the early twenties, though, cigarette sales began to outstrip cigar sales, and venders for singles or packs were made for the new market (p. 65).

Postage machines were also examples of early coin-operated vending (p. 66). They were a necessary convenience for dispensing stamps and postcards outside the post office.

Public toilets took advantage of a captive audience to sell assorted "necessaries" for personal needs (p. 67). Some oddities — collar buttons, pencils, and perfumes — were also found to be practical vending items. Almost every type of merchandise, at one time or another, has been sold through a machine.

As early as 1891, the success and proliferation of machines offering so many items and services raised questions as to their threat to "honest labor." But the vending machine has established a niche for itself as the most efficient way to sell a particular kind of product. In addition to its manageable size, almost everything sold from a vending machine is used on a regular basis and used up — eaten, drunk, smoked, or in some other way disposed of — within a short period of time.

In addition to their function of dispensing merchandise, vending machines also had to sell and advertise their wares, and employed a variety of methods to do so: Some have brightly colored exteriors or decals, or gleaming chrome or mirrors. Others use bold lettering or bizarre shapes to attract. An extensive use of glass exposes both the wares and the mechanism. Some cabinets successfully rely on quiet dignity; others, beauty. Intriguing animated figures and animals attract the buyer as much as the product itself.

Today's vending machines (alas!) are rarely more than functional boxes. Mass production and a general streamlining of design have eliminated the more playful aspects of vending, particularly the old, charming countertops.

Antique vending machines have generally ranged in price from $25 for the common gumball machine to $500 for the rarer collector specimens. Some of the rarest and most desirable pieces, like the animated venders and the glass venders, sell for more.

NIAGARA

Folk art and mechanical design are superbly united in this amusing and rare vender. Handcrafted and signed "Build by Otto Schmidt," the Niagara conceals a clockwork motor in its 27″ oak cabinet. At the drop of a penny, the boy twice tips his hat and waves his cane as the gum is ejected from the mouth of his comical friend. The automaton's glass eyes are set in plaster, and the large head is made of iron. Produced by the Niagara Pepsin Gum Co. of Buffalo and patented Aug. 23, 1904.

TEDDY BEAR

The cuddly bear became the rage after Teddy Roosevelt's famous encounter with a bear cub. This motif was adopted by the Caille Brothers Co. for a new vending machine introduced around 1909. For a penny, Teddy makes a complete somersault as the machine delivers a stick of gum. One winding of the clockwork motor is sufficient for 130 lively performances. The floral cast-iron front frames the gum and a charming lithographed scene of storybook animals.

BIRDIES GUM VENDER

The two birds chirp, move their heads and beaks, and flip their tails while dispensing a gumball. Their song comes from hidden bellows and their animation from a clockwork motor. Made by Purdy-Moore Mfg. Corp. c1928.

AUTOMATIC CHICKEN

Another auditory vender, this embossed tin chicken cackles as she "lays" a real hard-boiled egg or a confection-filled tin egg for a nickel. Many variations were made of this motif. c1905.

BULL'S HEAD PERFUME

Taking the bull by the horns causes him to snort perfume from his nostrils. The penny perfume vender is made of cast-iron by the Continental Novelty Co. and was patented Dec. 27, 1904. A similar version was built in Canada.

HAPPY JAP

Most happy is he when given a penny! A windup motor responds by pushing a gum through his teeth. The 13" cast-iron head was made by the White Vending Machine Co. and patented Aug. 25, 1902.

BLINKEY EYE

A push on the rod awakens the sleeping face that watches the patron receive a stick of gum and then shuts its eyes again. Made of cast-iron by J.W. Hange Foundry Co. Pat. Nov. 26, 1907.

SMILIN' SAM

...from Alabam', The Salted Peanut Man. Pulling his tongue vends a handful of peanuts. Made of cast aluminum by the General Mdse. Co. in 1931, it was copied again in 1975.

PULVER'S CONFECTIONS

Inside the richly illustrated cabinet of embossed and lithographed tin, various obliging figures such as The Professor or Buster Brown "help" deliver the chocolates or gum. Made by Pulver Mfg. Co. 1902.

PULVER'S GUM

The Yellow Kid pivots towards the gum as it drops down the chute. The popularity of their first windup animated vender led to Pulver's long-lived commercial success. Pulver Mfg. Co. Pat. Dec. 29, 1897.

COP AND HOBO

Officer Casey gleefully bops the hobo with his nightstick as the hobo's eyes and tongue flap in astonishment. Pulver Mfg. Co. continued to produce figural gum venders in bright porcelain cases through the early 1950s. c1930.

LADY PERFUME

In her garden, the lovely papier-mâché maiden automatically sprays perfume from the flowers on her bosom. A miniature of Mills Novelty Co.'s life-size arcade piece, both machines were great favorites with the ladies. 1905.

PERFUME SPRAYER

Perfume your handkerchief. Cupids offer a selection of four fragrances at a penny per spray from the colorful cast-iron front. Patented by Mills Novelty Co. July 11, 1916.

BAKER BOY

With the turn of the crank, the baker turns to the oven, opens the door, scoops up a gumball, and nodding, turns to send it down the chute. Many variations of this popular collectible have been made. Manikin Vendor Co. 1928.

E-Z

Also a trade stimulator, this 5¢ vender sells hollow gumballs containing numbered papers good for prizes or cash awards. Note the footed iron base. Ad-Lee Novelty Co. c1930.

ADVANCE

Operators appreciated its reliable mechanism, so the Model D was around many years. And penny gumball venders still offer the cheapest treat today. Made by Advance Mfg. Co. in 1923.

COLUMBUS

No shop could afford to be without a gumball vender, and this classic shape was one of the most popular. Made of enameled iron, it could also vend nuts or candy. Columbus Vending Co. c1927.

SUPREME

Since the Supreme alternatel vends one, two, or three bal the prudent buyer bought si for 3¢ and the lucky buyer coul get three for 1¢. Appleton Nove ty Co. 1927.

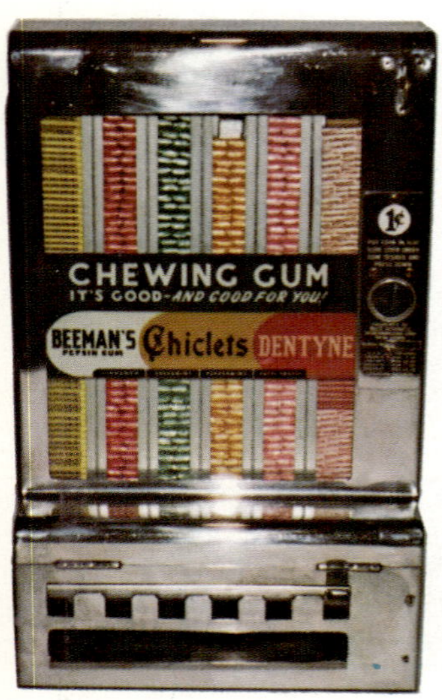

CHEWING GUM

Often associated with NYC subways, thousands of these machines vended cellophane-wrapped two-chiclet packages in other locations as well. Mills Sales Co. Pat. Dec. 22, 1936.

MASTER

Although only produced for six years, Norris Mfg. Co. turned out a large quantity of this adaptable machine. One penny buys one measure and a nickel buys five. Pat. Aug. 14, 1923.

MINT VENDER

A nickel in the top of one of the three columns automatically releases a roll of mints. The lithographed paper front makes this an irresistible machine. O.D. Jennings & Co. 1932.

ZENO GUM

This porcelain version of Zen Mfg. Co.'s earlier wooden cab net still employs a clockwoun mechanism for instant deliver Pat. May 5, 1908.

ZENO GUM

A 5¢ package of gum is automatically pushed into the chute by the visible clockwork mechanism. Zeno Mfg. sold unusual flavors such as yucca and vanilla cream. Pat. July 15, 1902.

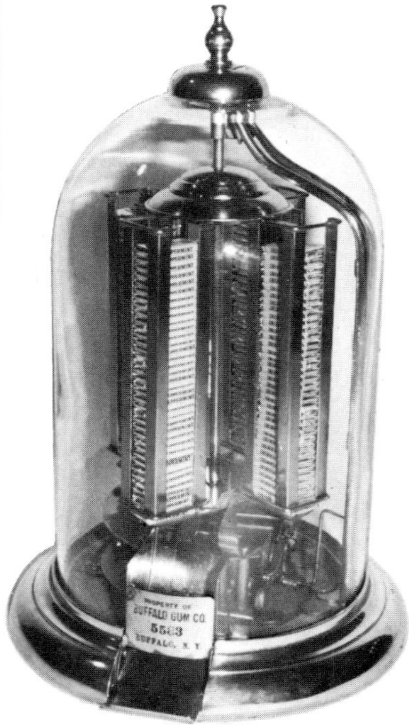

BUFFALO GUM

Inside the bell jar, a windup motor rotates the brass hoppers to dispense a tab gum. A penny dropped in the top starts the action. Buffalo Gum Co. Pat. June 4, 1907.

TRUE BLUE GUM

Glass-domed machines are fragile, so few have survived. Here, double columns offer a choice of two flavors. True Blue Gum Co. 1906.

CHICLETS GUM

The bright green label catches the eye and invites a sale. With a different decal, the same machine could sell chocolates, as on p. 63. National Vending Machine Co. 1907.

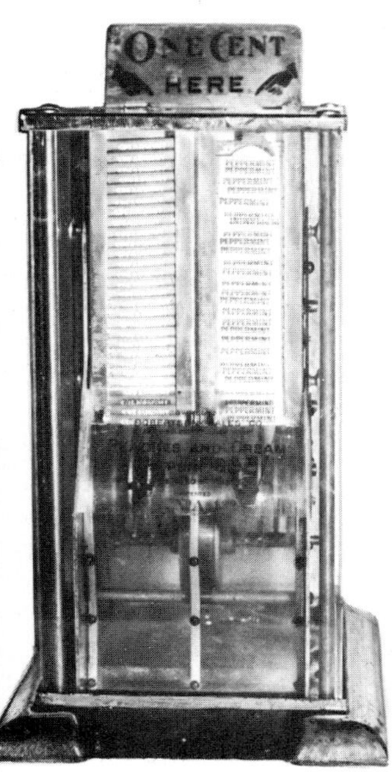

PEACHES & CREAM

Exposed mechanisms entertain while working. A spring motor drives a wheel that catches the penny that pushes out the gum. Operated by the Robertson Sales Co. 1906.

PEPSIN GUM

With the characteristic whir of a clockwork motor, a tab gum is delicately delivered. This machine is 10" high with a wooden top and base. Automatic Sales Co. Pat. Nov. 24, 1903.

COLGAN'S GUM

A clockwork motor pushes out Taffy Tolu from alternate chutes. The fifth penny is rewarded with a gum from both sides. National Vending Machine Co. Pat. June 24, 1902.

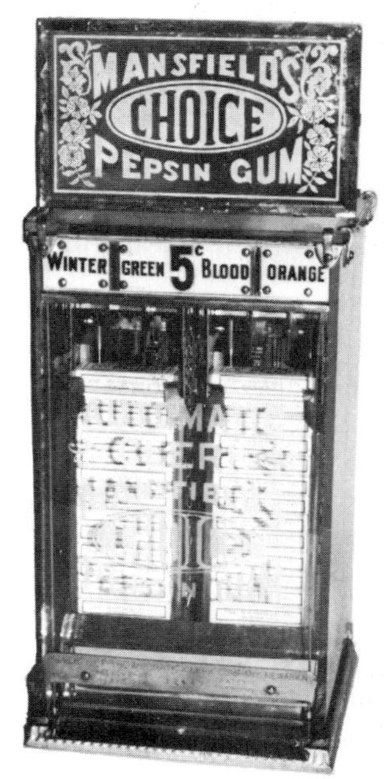

MANSFIELD'S GUM

The Automatic Clerk Co. made almost 100,000 units of this etched front glass classic. When a sale is made, the bell rings and a gum is flipped off the top. Pat. Nov. 19, 1901.

ADAMS' TUTTI-FRUTTI

Machines such as this were widely and successfully operated by Thomas Adams. Their convenience greatly influenced the American public to chew gum. Railway Automatic Sales Co. 1898.

ADAMS' PEPSIN GUM

Its large capacity requires less frequent restocking, so this machine's style was used by many venders placed in high-traffic areas. United States Slot Machine Co. Pat. Dec. 4, 1894.

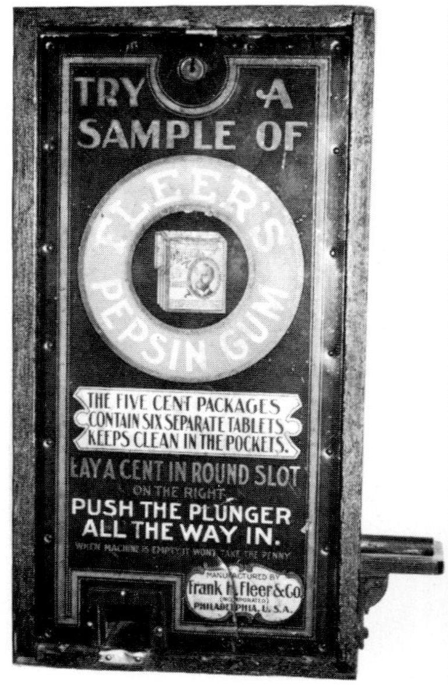

FLEER'S PEPSIN GUM

Frank H. Fleer & Co. Inc., of today's Double-Bubble fame, designed this machine to vend a sample stick of their gum as incentive to buy the 5¢ pkg. pictured on the front. c1898.

ZENO GUM

Pineapple? Lemonade? Zeno Mfg's earliest gum vender is made of oak, and employs a clockwork motor to gently push out a stick of their unusually flavored gums. Pat. Aug. 1, 1893.

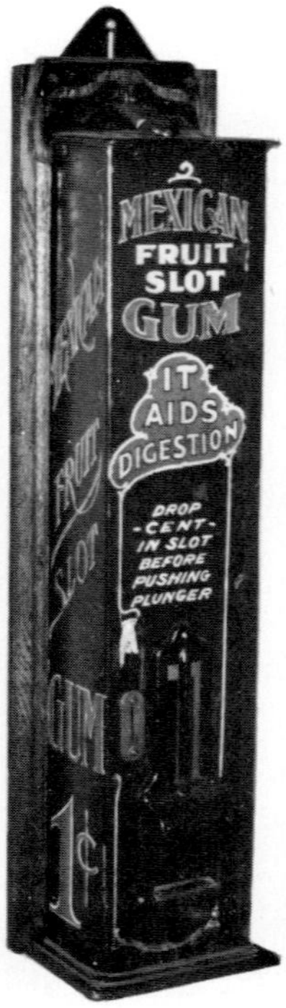

MEXICAN FRUIT

Graceful lettering adorns the front and sides of this colorful blue and red embossed and lithographed tin gum vender. London Novelty Co. c1900.

BLACK DIAMOND GUM

"We chew Black Diamond Pepsin Gum. We chew it cause most others are bum. We chew it now and we want some more, to chew until we can chew no more." Bethlehem Mfg. c1900.

PANSY GUM

Pansy Gum comes wrapped in a fortune and love letter, and is vended from the blue and red art nouveau cabinet. London Novelty Co. c1900.

GOO GOO GUM

The oxidized-copper iron front has no lettering except the "one cent" under the coin chute, allowing the operator to stencil his brand on the pine sides. J.W. Hange Foundry Co. 1899.

STOLLWERCK CANDIES

Gold-lettered red glass plates are set in an oak cabinet with a blue and white porcelain marquee. Four columns vend caramel, gum, and the popular chocolate. Volkmann, Stollwerck & Co. Pat. Sept. 27, 1892.

STOLLWERCK SAMPLER

Fine walnut with a delicate brass trim encloses a clockwork motor to turn a vertical wire corkscrew that drops out a 5¢ chocolate bar with each half revolution. Schilling-Stollwerck & Co. c1905.

WILBUR'S CHOCOLATE

A large chocolate producer, Wilbur used various machines for different locations. This lilac-decaled vender was clearly made for indoors. National Vending Machine Co. Pat. Nov. 1, 1904.

WILBUR'S SWEETS

Designed for heavier use, this red and white porcelain cabinet is topped with Wilbur's familiar trademark of Cupid with a cup of cocoa. The same machine also vends gum. 1899.

SUNNY CAL

This vender is unusual in that it is handpainted and shaped like the product it vends. The inner tray advances a 10¢ slice of fruitcake into the opening. R.O. Whyman & Co. Inc. c1930.

OLD MILL

Slum (the trade word for hard candies and novelties) is pushed into the chute by the windmill's arms. The painted Dutch motif completes the theme. International Mutoscope Co. 1928.

EVER READY LUNCH

A counter-sized automat dispenses various foods at 10¢ per selection. Lacking refrigeration, this was probably not a longrun success. c1930.

ENTERPRISE PEANUTS

The simple oak vender with its tempting display window sells salted Spanish peanuts by the box, via a clockwork motor. Enterprise Vending Machine Co. c1905.

ORBIT, JR.

Tin ads on the sides promise Mills Novelty Co.'s own brand of Spanish peanuts is "good for digestion." The claw-footed iron cabinet is embellished with a leaf motif and inset with beveled glass. 1902.

NUT & CANDY VENDER

Dragons in relief on the antiqued-copper iron front guard the contents of Freeport Novelty Co.'s attractive vender. Wooden sides have round windows and eagle and flag decals. 1905.

CHICAGO

The directions cast in the spool-trimmed iron base read "Push penny in slot, Pull handle over." Note the unusually shaped glass globe. Coleman Hardware Co. Pat. Sept. 8, 1908.

CONFECTIONS VENDER

Berger was a prolific inventor of mechanical devices. This elaborate example of his genius has a cast front, painted metal sides, and demonic supports. Paul E. Berger Mfg. Pat. June 21, 1904.

CLIMAX 10

One of the earliest venders using a glass globe to expose its wares, the Climax 10 with its footed iron base was long a commercial success for Advance Machine. Pat. Feb. 11, 1902.

PEANUT VENDER

The balloon-shaped globe and pin-striped iron base give a smart appearance. Pulling the drawerlike handle releases a shower of peanuts. Vending Machine Co. of America. 1910.

THE LEEBOLD

Even in the mid-1920s, the Victorian styling of The Leebold's florid cast-aluminum top and base was still admired. C.E. Leebold. c1925.

LITTLE PERFECTION

Mills Novelty Co. advertised their perfect peanut vender as sporting the new sanitary dust-proof chute and cup, and stronger salt and weatherproof mechanism. Pat. Sept. 10, 1912.

WEEKS PENCIL VENDER

By positioning the letter slides beforehand, the patron can imprint his name on a pencil. Early models like this were made of oak, but later versions of cast metal operated into the 1950s. Charles Weeks Co. 1923.

COLLAR BUTTONS

The insertion of a nickel releases a single collar button and rings a bell to announce the sale. Columns rotate inside the glass walls for style selection. Price Collar Button Machine Co. c1905.

COLLAR STUDS

Collar buttons were once a necessary part of a gentleman's attire, used to secure his removable collar to a not often laundered shirt. Here, a clockwork machine automatically vends a choice of buttons for a nickel. c1901.

HOWARD HOME HUMIDOR

Placed in elegant hotel lobbies and men's clubs, this seemingly innocent magazine table actually vends a pack of cigarettes for 15¢. C.B. Howard Co., Inc. produced a variety of table shapes concealing cigarette machines. 1931.

LITE-O

Made to resemble a gas pump, the Lite-O Filling Station is an "honor" vender. A patron could pump lighter fluid from the nozzle to fill all of his friends' lighters until the handle is released. Brearley Ltd. 1930.

1¢ LUCKY STRIKE

In 1931, long before Lucky Strike green went off to war, those who lacked fifteen cents for a pack of cigarettes could buy a "loosey" for a penny. Wilson Mfg.

PENNY POSTCARDS

With its eagle-topped dome and cupid medallions, this vender might have come from Ruritania, but was actually made by the U.S. Postal Advertising Co. A picture postcard, like the sample displayed in the dome, was obtained by a turn of the crank. c1900.

MONFORT'S POSTAGE

A two-cent stamp and a penny envelope are more easily vended for a single nickel, so the above postage clerk obligingly returns two cents change. Monfort Stamp and Envelope Co. Pat. Jan. 10, 1893.

U.S. STAMPS

The official-looking shape and patriotic red, white, and blue coloring give the impression of government sponsorship, but this stamp vender was actually made by Mills Novelty Co. for the American Vending Machine Co. in 1915.

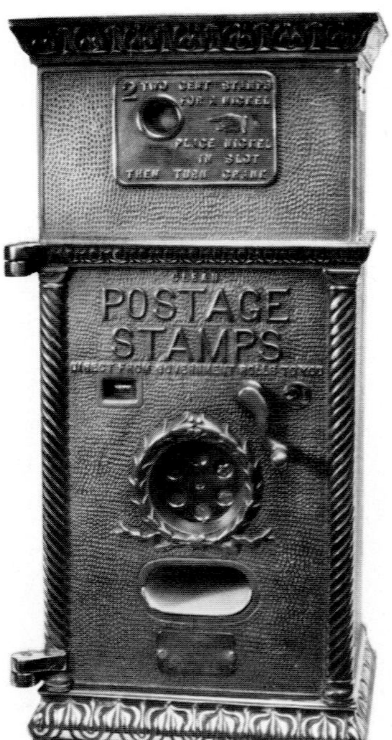

POSTAGE STAMPS

"Clean. Direct from Government rolls to you" is cast on the face of this all brass stamp vender. Schermack was the first company given official sanction in 1907 by the U.S. government for selling postage stamps from a coil. 1922.

KONE-KLUTCH

For a nickel, the patron has a choice of chopping off four 1¢ or two 2¢ stamps when he presses the handle. The beveled glass shows the inner mechanism to imply that the stamps are germ-free. Kone-Klutch Corp. Pat. July 11, 1911.

POSTAGE STAMPS

An elaborate turreted iron front gives this postage vender a stately presence and provides a built-in sponge for moistening stamps. Stamp packets were vended before the advent of coiled rolls. National Vending Machine Co. c1905.

CHUM RAZOR BLADES

The revolution in shaving equipment made disposable blades an inevitable vending item. Made of cast-iron and just 5," this is the smallest machine known. Operated by A.S.L. Sales Co. c1928.

THE AUTO DOCTOR

With a perfect bedside manner, The Auto Doctor offers cold tablets, cough troches, and gum to cure headaches, dyspepsia, and halitosis. Etched and beveled glass front. Auto Doctor Co. 1892.

PILL VENDER

Pulling the knob turns the notched wheel to allow the patron to remove two medicinal aids from the opening in the glass top. Mfr. unknown. c1895.

JERGENS LOTION

One cent buys a sample squirt of lotion. Satisfaction could lead to store purchases. The chic art deco styling is done in black cast-iron and glass. Lotion Dispenser Corp. c1934.

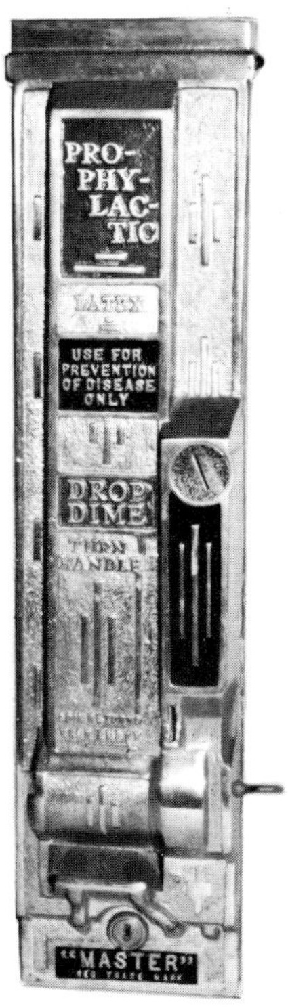

PROPHYLACTICS

Great tact is needed to vend this product. Disease prevention is emphasized and side benefits prudently ignored. Made of cast aluminum to hang in men's rooms. Norris Mfg. Co. c1930.

FEMS

When women's sanitary supplies were introduced, vending machines were quick to offer a discreet place of purchase. A pastel lady adorns the porcelain case. Hospital Specialty Co. c1925.

TOILET PAPER

Pity the poor patron without a penny! Even in the privy, the proprietor made a profit. Made by American Vending Machine Co. and patented Nov. 3, 1903.

TOWEL AND SOAP

A mechanical bathroom attendant, with a bright blue and white porcelain front, offers the necessary toilet equipment for a nickel. American Vending Machine Co. Pat. July 5, 1904.

THE "ELM CITY"

The New Haven Car Register Co. had to obtain approval from the Internal Revenue Dept. to sell cigars individually. Arched windows display a choice of brands beneath the pediment carved in the popular Eastlake style. Pat. Dec. 17, 1889.

THE VANDIVER CIGAR VENDER

Here's some zany action: Insert nickel. A bell rings. Slide the plunger along the top track until you reach your favorite brand. Press down on plunger. A needle stabs the cigar. Draw the plunger back. Cigar will fall into the tray. If you are a good marksman, your cigar should still be smokable. Star Novelty Mfg. Co. Pat. Dec. 7, 1909.

THE HONEST CLERK

This sturdy and elaborately scrolled salesman earned his space and did not need the storekeeper's watchful eye. The name indicates that the problem of the untrustworthy clerk is solved. Jackson Supply Co. c1908.

POPE CIGAR VENDER

Everything is visible to assure that the product is sanitary and properly preserved. A dime starts the clockwork mechanism to push the cigars until one catches and spills out on the tray. Pope Automatic Co. Pat. Jan. 19, 1909.

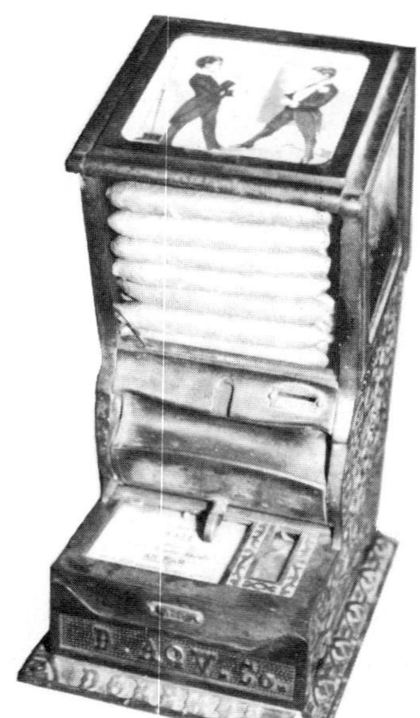

DOREMUS CIGARS

The odd shape here is actually functional. The slanted top displays an ad and the center section gravity feeds the cigars to the trough. This machine was also available with a cigar cutter on the side Doremus A.V. Co. c1905.

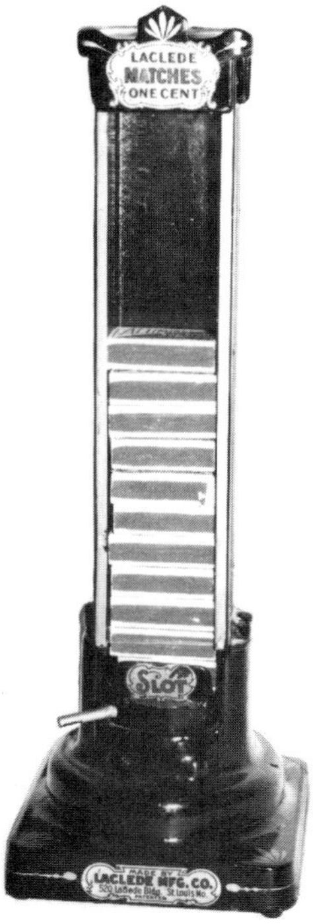

LACLEDE MATCHES

A flip of the lever sends a spring-loaded box of matches soaring. Green decals and gold trim on black enameled iron give a graceful classic appearance. Laclede Mfg. Co. c1915.

ONE CENT MATCHES

The glass dome displays rotating columns of matchboxes atop a footed cast-iron base. A box is dispensed when the handle is slid forward. Advance Machine Co. Pat. Feb. 20, 1916.

1¢ MATCHBOXES

Griffinlike creatures guard the chute from which the lever releases matchboxes from alternate columns. Operated by the Philadelphia Vending Co. 1912.

PENNY MATCHES

Northwestern Vending Corp., still in business today, introduced their first machine in 1911. A patron is offered a free light and use of the cigar cutter, but a box cost a penny.

SAFETY MATCHES

A glass dome houses a yellow fringed cone atop a four-columned carousel that revolves as the rod is pushed. Jackson Vending Co. c1905.

DIAMOND MATCHES

The dial on the front could be changed to vend one to four matchbooks for a penny, according to the storekeeper's discretion. Beaton & Caldwell Co. c1915.

MATCH VENDER

Mills Novelty Co. claimed "96 out of 100 men smoke," so match venders were essential for the shopkeeper. Cast-iron and oak construction made this a dependable fixture for years. 1911.

KNAPSACK MATCHES

An ingenious clockwork motor nudges a brass-plated match tin onto the tray. When the column is empty, the next automatically revolves into place. United Machine & Supply Co. c1902.

Service Machines

Service machines are vending machines that sell the performance of a specific activity for a limited period of time. Like vending machines, they can be devised for an almost unlimited number of functions, provided the necessary conditions are observed: The location must suit the function for which the machine was designed. Activating the service itself must be simple — usually a single on-off or open-shut operation. The service must be popularly and frequently needed, and must be performed within a finite and easily regulated amount of time. Given these ingredients, almost any mechanical service can be (and has been) adapted to coin operation. Unlike other vending machines, and with certain exceptions, most service machines do not replace a servant or attendant, but perform what no single human being could.

The application of the coin-in-the-slot principle yielded some strange examples, some of which are shown on these pages. But there are others: A hat and coat rack, patented on August 19, 1913, required a penny to retrieve one's property. A toilet seat, patented August 10, 1915, would not lower to a usable level until a penny was deposited. And coin-controlled gas meters were commonly found in rooming houses.

One major category of service machines measures and records specific personal attributes, always a source of endless interest to the consumer. One of the earliest and most successful of this type was the coin-operated scale. As competition developed, manufacturers added many unusual novelties to the original function, such as music boxes, chimes, skill games, talking features, penny-backs, fortunes, printed weight tickets, and vending attachments. Other machines measured height and even eyesight.

Turnstiles and door locks were also among the earliest machine designs produced in the mid-1880s, and are still with us today.

Any electrical appliance could be fitted with a coin-operated timer to become a service machine. On the theory that it was preferable to sell a service rather than give it away, hotel owners, for example, sold the use of ordinary electrical devices — radios, room heaters, bed vibrators, and other amenities — adapted to coin-operated timers.

Service machines quickly became an indispensable part of the modern economy. The merchant who bought one of these had the advantage of operating a machine that needed no stocking and little attention, and a captive public who either needed its service or was intrigued by its novelty.

DICTIONARY

The spring mechanism keeping the lexicon closed is released by a penny. The reader can look up numerous words by holding the book open. When released, it snaps shut. Coin Operating Book Holder Co. Pat. Mar. 6, 1900.

LAMP

A penny is inserted in the bottom of the base, and the lamp stays lit until it is shut off. It is an odd combination of an art-nouveau shade and baroque-style brass base and finial. Patterson Studies Inc. Pat. Dec. 1926.

PREPAYMENT FAN

Five cents buys an hour's worth of breeze. This standard model has brass blades inside a simple wire cage atop a cast-iron base, and was probably operated in hotel rooms. Made by G.E. for Electric Fan Motors Co. Pat. Feb. 7, 1905.

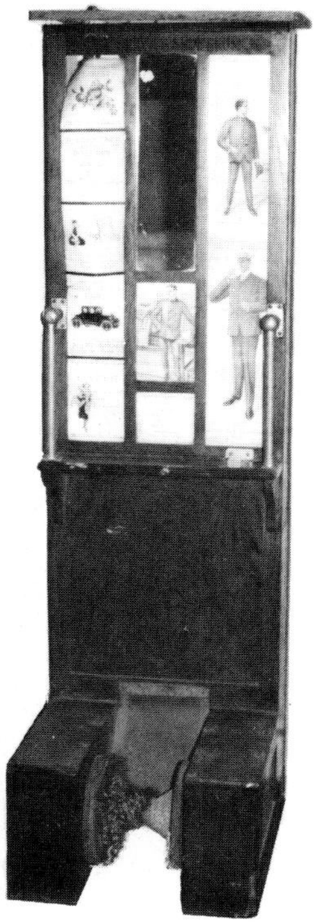

SHOE BRUSHER

The patron grasps the handrails and gazes at his reflection in the beveled mirror or reads the illuminated ads while horsehair bristles brush his shoes. National Service Enterprises 1912.

READ YOUR HEIGHT

After leveling the tops of his ears with the center of his eyes while sighting into the etched lines on the mirror, the patron inserts a penny. Windows open to reveal height. Slezak Co. c1925.

SHOE BUFFER

Porcelain signs on this oddly shaped bright orange and black cast-iron shoe buffer promise to restore the shine without adding "dressing." Automatic Machine Co. of Rhode Is. 1914.

TALKING SCALE

Pressure on the treadle moves a phonograph needle to a vertical record that squeaks out the weight. A lucky weigher can even get his penny back. Notice the sylvan scene handpainted on the metal front. c1915.

WATLING SCALE

By correctly predialing his weight, a good guesser could get his penny back. The exposed mechanism is housed in a dignified oak cabinet. Watling Mfg. 1902.

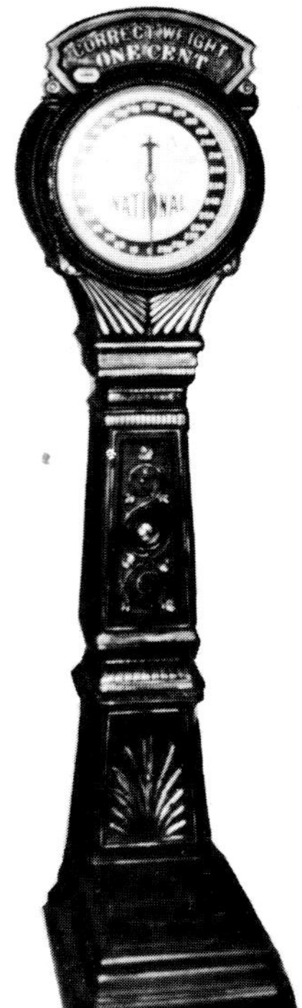

THE NATIONAL

National Novelty Co. produced many elegant scales. This classic example has a porcelain face set in sculptured cast-iron. 1910.

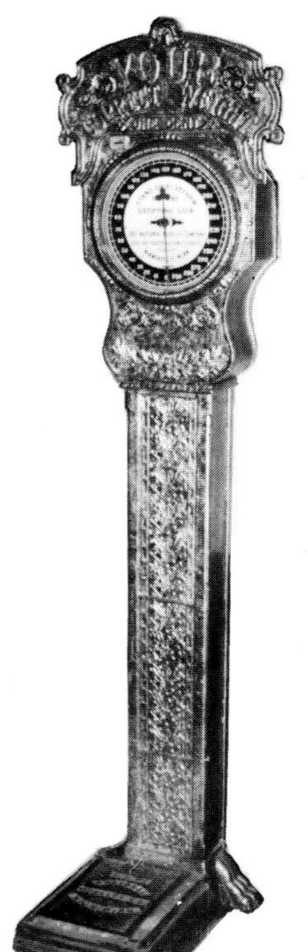

NATIONAL WEIGHER

This gilded scale is highly ornate and stands on clawed feet. The colorful blue and white porcelain face registers up to 300 lbs. Made by the National Novelty Co. 1910.

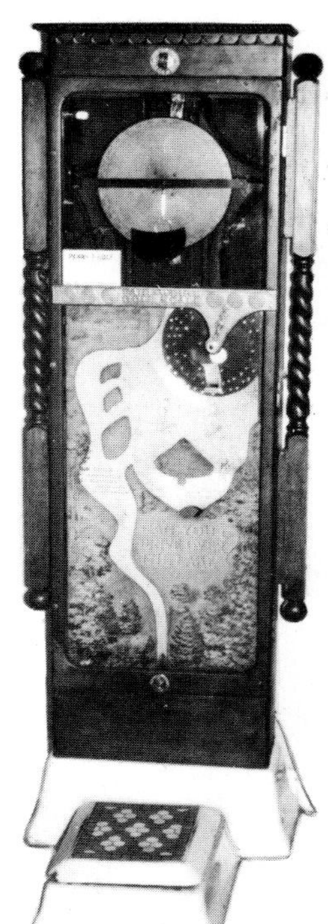

FAIR WEIGH GOLF

Turn the knob to kick the penny onto the fairway. The skilled player gets a free weigh if his penny drops through one of the four exits into the cup below. Colonial Novelty Co. c1920.

ARCADE

Some form of diversion to stimulate the imagination and provide escape from everyday cares is essential to the human psyche. It naturally follows, then, that diversion would be offered as a commodity for sale wherever a sufficient number of people could be gathered to make a commercial enterprise profitable. Until the late nineteenth century this limitation meant that traditional forms of the entertainment business — plays, concerts, circuses, athletic contests, and pleasure gardens — could be supported on a permanent basis only in larger cities. People in small towns had to wait for traveling companies or fairs, and between times rely on homemade amusements or limited local talent.

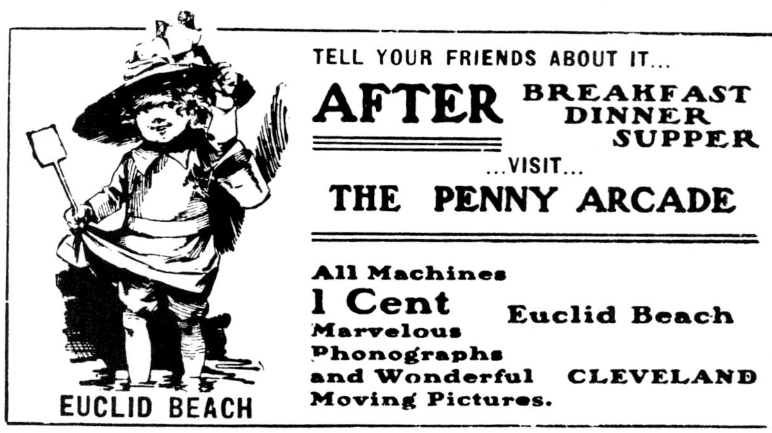

It is hardly surprising, then, that the appearance of coin-operated Edison phonograph parlors on America's Main Streets at the turn of the century should have proved to be a popular new form of diversion. To provide novelty and keep the customers coming, other kinds of machines — kinetoscopes, coin-operated adaptations of punching bags and strength testers from the local saloon — were added. The Edison Parlor became the Penny Arcade, the Penny Wonderland, and the Automatic Vaudeville.

This new kind of business was so lucrative that when trolley car companies began setting up amusement parks at the ends of trolley lines to encourage evening and weekend ridership, they invariably included Penny Arcades as one of the attractions. And at the St. Louis Exhibition of 1904, Mills Novelty Company touted the money-making potential of the coin-operated amusement machine with its "Edisonia" housed in a brilliantly illuminated "Spectatorium." Caille Brothers also had a similar display. One of the selling features was the suitability of this new entertainment for the entire family, although the best customers were still the young male "sports."

Thus within a few years, the arcade, according to H. S. Mills, had become a permanent institution, "as important as the theatre, the opera, and circus." Every city of at least 10,000 population, he claimed, could support an arcade. And indeed, up until World War I, the pennies added up to respectable profits. During this early period the classic penny arcade machine types were created.

Peep shows constituted a major category of amusement devices (pp. 80, 88). Using slides, stereoscope cards and film strips (some even combined with music), they offered scenes of the world's wonders, comic vignettes of daily life, news events, sports, and mildly titillating views of lovely ladies. One of the most successful viewers was the Mutoscope (p. 80), which gave the illusion of motion as the crank was turned. The only company to produce this machine was founded in 1895 by the Rabkin family and was first known as the American Mutoscope and Biograph Company. (Later, about 1920, the name was changed to the International Mutoscope Reel Corporation and the products expanded to include the famous "4 Poses for 10¢" photograph machines. In 1930, the current name, International Mutoscope Corporation, was adopted.)

Strength testers (pp. 77, 78, 80, 82, 83, 88) came in an extraordinary variety of styles, testing an equally extraordinary variety of muscles. There were over thirty different methods for men and boys to either compete among one another or to impress the weaker sex.

Fortune tellers (pp. 77, 79, 82, 86) were indispensable fixtures in amusement arcades, because although no one wholly believed in their predictions, most people half-believed, at least for the moment, in their mysterious powers. Automatons in the form of gypsies, wise old ladies and anthropomorphic animals had (and have) an especially irrestible fascination, and are prized collector's items today.

Countertop models of electricity machines (p. 89) were primarily advertised as cure-alls, but the floor models (p. 81) in arcades were used mainly as competitive tests of endurance. Watching the reaction of the patron to the charge of electricity was amusing to the onlooker as well and furnished an additional source of entertainment for the arcade crowd.

Novelty venders were natural adjuncts to amusement machines. Personalized souvenir tokens (p. 82) were always popular, and card machines selling love letters (p. 80) and picture postcards appealed to the collecting instinct by offering various series of cards.

Whatever the type, few could resist trying each kind of machine in a typical arcade. Besides arcades, two or three of these machines could also be found in other locations where the public congregated, such as poolrooms, bowling alleys, restaurants, and hotel lobbies, to provide extra income. By 1913, a supplier's catalog could confidently state that the coin-operated amusement arcade was "the only business where failure is absolutely impossible."

After World War I, however, failure became increasingly possible. There was competition from other forms of mechanically reproduced entertainment — moving pictures, radio, and home phonographs. The doughboy who had seen Paree was not as easily beguiled. Lung testers had already been banned as unsanitary, and electric shockers had lost their therapeutic value. Prohibition had eliminated the saloons and restaurant bars where the men about town had tested their grips.

By the late twenties, a major producer, Caille Brothers,

Entertains More People at One Time Than Any Other Concession

Above photograph shows our 90 machine equipment, consisting of all the new highest grade large floor machines, designed for permanent locations such as parks, beaches, and penny arcades. There is no better variety or quality obtainable. This is the "last word" in penny arcade equipment.

Above photograph shows our 90 machine, large size standard equipment, consisting of large floor model machines, designed for permanent locations, such as parks, beaches and penny arcades. There is no better variety or quality obtainable. This equipment includes all our new models: 1 Humpty Dumpty Amusement machine; 1 Lindbergh Aviation Striker; 1 Atlas Dumbbell; 1 Love Tester, which registers the sex appeal; 1 Electric Enerziger; 1 Champion Muscle Developer Punching Bag; 1 High Striker; 1 Grandfather's Clock; 1 Hercules Ball Gripper; 1 Pull the Tiger's Tail Athletic; 1 Lighthouse Siren Athletic; 1 Striking Clock Athletic Squeezer; 1 Combination Gripper and Lifter; 1 Jackass Lifter; 1 Iron Claw; 8 Floor Model Photoscopes, electric model picture machine, with 5 sets of pictures in each: 1 Egyptian Seeress with 11,000 cards; 1 Crystal Gazer with 11,000 cards; 1 Cupid's Arrow on pedestal; 1 Electric Vitalizer; 1 Enchantress with 7,000 5c cards; 1 12-slot Horoscope Machine with 20,000 cards; 1 12-slot Whom to Marry with 20,000 cards; 1 Rajah Mystic Crystal Fortune Teller; 1 Jazz Baby Shimmy Shaker; 1 12-slot Eye Fortune Machine with 20,000 cards; 1 Palmistry machine with 12,500 cards; 10 Model D Card Venders, with latest steel mechanism, equipped with flashy signs, and including 25,000 cards from latest releases; 1 Cupid's Postoffice with 12,500 cards; 2 Play Ball Machines; 2 Lucky Horseshoe Penny Back Fortune Tellers; 4 Oracles (Fortune, Past, Present and Future); 2 Radio Love Message Machines with 5,000 cards; 4 Little Gypsy Fortune Tellers (1 of each kind); 2 Ball Gum Fortune Tellers (1 boy, 1 girl); 2 Hindoo Fortune Tellers (one of each kind); 2 Junior Dice Machines; 2 Kromo Kolored Kubes; 2 Bowling Alleys with special flashy display sign; 2 Five Ball Shooters with flashy sign; 1 Sambo Fortune Teller; 1 Smiling Joe Penny Back Ball Gum Machine; 2 Electric Shock Machines; 1 Counter Arcade Dice Machine; 4 Lucky Photo Venders with 2,000 5c cards; 1 Gold Mine with 5,000 5c cards; 1 Automatic Pistol Range (leased with equipment for life of our patents and to be operated only in penny arcade); 1 Bull's Eye Shooter; 1 Batter-Up Baseball Machine; 1 Golden Bell Penny Back Skill Game; 1 12-slot Astrology Machine with 20,000 cards; 1 12-slot Hair Fortune Teller with 20,000 cards; 1 Bell Jack; 1 Knockout Punch Tester; 1 Hole in One 5c Skill Machine. Machines placed in four rows as per above photograph, minimum floor space 25x50 feet; maximum space 35x60 feet. Net weight, 7,000 lbs.

90 Machines $9000

Exhibit Supply Company advertised this complete arcade in the early 1930s.

The interior of a typical Mills Novelty Company's Penny Arcade around 1905.

A Few Fine Arcades

MILLS EDISONIA
at the St. Louis World's Fair

MILLS ARCADIA
Memphis, Tenn.

MILLS EDISONIA CHICAGO
One of our Arcades, located at 278 State Street, Chicago

AUTOMATIC VAUDEVILLE
14th St., New York City

MERRYMEETING

Amusement Parlor!

A GOOD PLACE FOR THE OLD, MIDDLE AGED AND YOUNG TO DRIVE AWAY CARE AND HAVE A GOOD TIME. COME AND ENJOY YOURSELF.

We have the Following Attractions:

The Up-to-date Shooting Gallery,
Philippine Novelty Game,
Mexican Cane and Knife Board,
Fremont's Prize Shooting Gallery,
Also Punching, Lifting, Grip,
Muscle, Lung, Electric, Cigar
and Picture Machines.

Prizes Given Away Each Week

FOR THE BEST SHOOTING.

We use the FAMOUS

Match-maker Brand Cigars,

the BEST CIGAR ever given away in a place of this kind.

We are open for Business each day, from 8 A. M. to 10 P. M.

For Fun and Amusement this is the place. Come and see the finest Amusement Parlors in the State.

W. S. FREMONT

MANAGER.

N. B.—We buy and sell show property of all kinds, Tents, Etc. If you have anything to Sell, or want to Buy or Trade anything in that line, it will pay you to call on me.

had dropped out of the amusement business, and Mills Novelty was no longer the leader in the field. The Exhibit Supply Company, mainly by virtue of its Iron Claw digger machine (p. 81), one of the all-time successful arcade machines, went on to become the world's largest supplier of arcade equipment. Skill games, especially those based on popular sports (p. 84), manufactured by the Chester–Pollard Amusement Company, were added to the standard arcade lineup. Nevertheless, in spite of these innovations, arcade machines survived only in amusement parks.

Paradoxically, the Depression of the thirties brought prosperity to the amusement machine industry. The penny amusement machine provided a cheap form of entertainment. Countertop machines (pp. 85, 86) reappeared in Main Street stores. The most successful, Whirl-a-ball, started a new fashion, and variations such as the Boop-a-Doop (p. 87) were everywhere. New companies sprang up to produce inexpensive penny machines. Pinball machines emerged as an important and separate division of the industry (p. 91). After the repeal of Prohibition, newly opened bars once again became an important market for coin machines. Jukeboxes, equipped with electronic amplifiers and multiple selection, became a necessary bar attraction, and so another branch of the industry was established.

Unfortunately, invention and development stopped in the forties, since production facilities of coin-op manufacturers were diverted to the war effort. By the time the country was adjusting to peace again, television had come along to dominate the entertainment business. For a while, it was feared that the customers would never leave their homes.

Nevertheless, in spite of the tough competition, the lure of the amusement machine is still hard to resist. Today, in restaurants, malls, and amusement parks, we can (for 25¢) destroy the Martian invader and play a game of Pong on an electronic screen or consult a computer to determine our bio-rhythm. If the penny arcade is no longer the poor man's club and a source of innocent merriment for the entire family, it is by no means dead. It is remarkable that the early inventors should have so accurately gauged the interests of the public. Nobody is ever too old or too sophisticated to pass up a chance to try his skill, to test his strength and reflexes, to compete in a mechanical version of a popular sport, to guide a digger for a trinket or consult an oracle for a prophecy about love or money, to collect a series of cards or have his name printed on a medallion.

The prices of antique arcade machines on the collector's market range widely according to rarity, function, and condition. Floor model lung testers and automaton fortune-tellers sell between $1000 and $5000, with some exceptionally rare examples going even higher. Strength testers, peep shows, and floor model electric shockers are found at prices between $500 and $2000. Countertop machines generally range between $200 and $1500. These prices apply to excellent original, or restored machines. Poor condition or unrestored pieces usually sell for less.

MADAME ZITA

Who would not like to know what destiny holds? Ask Madame Zita a question and her card will answer yes, no, or maybe and tell the future. In a fifteen-second sequence of movements, the automaton bows to her audience, opens the bin, picks up a fortune card, deposits it in the basket, and blows a farewell kiss. With her velvet hat, glass beads, and buckled shoes, Madame Zita is richly attired in gypsy style. This electric version was made about 1905, and Roover Brothers' all-mechanical model was patented Feb. 4, 1896.

UNCLE SAM

All patriotic Americans want to shake hands with Uncle Sam, so this classic strength tester was taking in pennies at arcades for many years. The firmness of the grip is indicated by the arrow on the dial, and a score of 300 rings the bell. Uncle Sam stands over six feet tall and is made of enameled iron mounted on a quartered oak cabinet. Copied by the International Mutoscope Co. in the 1920s, and privately in 1978, the machine was first produced by the Caille Brothers Company around 1908.

HAT BLOWER

"Get the crowd laughing and they'll stick." Mills Novelty Co. devised ingenious lung testers to attract and amuse the people. Testing lung capacity by blowing into a tube until one is red-faced amuses and also encourages competition. The idea of this massive ten-foot machine is to blow the hats off the comical papier-mâché heads by raising the center pointer past each one. Five hats blown off rings a bell, and if all eight heads are bared the lucky bag-of-wind gets his penny returned. 1904.

LION LUNG TESTER

It's thrilling to hear the lion roar and see its eyes light up when a champion breath holder blows into the tube. The second hand on the dial automatically indicates the length of time the contestant holds his breath. Although the most impressive of arcade machines, lung testers were banned and subsequently discarded around WWI as unhygienic. (Many mouths on the same hose helped spread tuberculosis and the common cold.) Because of this, lung testers are rare machines today. Made by Mills Novelty Co. in 1904.

ELEPHANT

The elephant, dressed in satin and lace, moves her head and arms while her trunk flips the pages of her book of predictions. When she lowers her trunk and stops, the patron may read the selected fortune. Roover Brothers Co. c1899.

THE EDUCATED DONKEY

Moving her head from side to side, the lady donkey surveys the crowd as she flips her baton to spin the wheel. The number at which it stops corresponds to a numbered fortune listed on the wall behind her. Roover Brothers Co. c1899.

PUSS IN BOOTS

The fabled cat in ruffled shirt and velvet breeches waves his baton, scoops up a fortune-telling card with the tray on his paw, and deposits it in the basket. Roover Brothers Co. c1897.

GRANDMA'S PREDICTIONS

Her eyes roll, her head nods, her chest heaves. The wise old grandmother passes her hands over the cards and stops at the proper fortune. A corresponding card falls into the tray. William Gent Mfg. 1929.

SIBILLE

The Queen of Hearts, adorned in exotic clothing and a jeweled diadem, reveals one's fate with playing cards. Bowing, she indicates the chosen fortune-telling hand as a printed portent simultaneously drops into the cup. Mills Novelty. Co. 1910.

PRINCESS DORALDINA

Her youth and beauty attract the arcade customer. A wax head, real human hair, and lifelike movements make the princess appear to be alive and give her fortunes believability. Doraldina Corp. 1928.

CUPID'S POST OFFICE

Affairs of the heart are a surefire theme for attracting the public. Here, Dan Cupid oversees the delivery of love letters for the lady or gent. Mills Novelty Co. 1904.

MASCOT

Two machines in one, the Mascot tests strength of both lungs and muscles. Charts list average scores for occupations from blacksmiths to wagonmakers. Caille Bros. 1904.

RED TOP LIFT

A comic face surmounts this seven-foot strength lifter. At 1000 lbs. a bell rings, calling attention to the player and challenging others to the test. Caille Bros. 1904.

TUG-O-WAR

"Have a pull with Mickey Finn." Tug the rope until the dial registers 750 lbs. A bell will ring and a devil's head appear behind Mickey. Caille Brothers. Design patent July 12, 1904.

CAIL-O-SCOPE

Comedy, travel, sex, or adventure are offered at a penny a show. 15 framed stereo views revolve past the lens on a wheel driven by a clockwork motor. Caille Bros. 1904.

MUTOSCOPE

Turning the crank on the front of the "Iron Horse" flips cards past the lens to produce the illusion of moving pictures. American Mutoscope Biograph Co. Pat. Nov. 5, 1895.

QUARTOSCOPE

This penny picture machine displays four different sets of twelve stereo card shows. The viewer has to spend four cents to see them all. Mills Novelty Co. Pat. Oct. 12, 1897.

ROSENFIELD

Glass slides in a stereo viewer and an Edison phonograph with earphones make this an early sight and sound attraction. Rosenfield Mfg. Co. 1906.

IRON CLAW

Turn the handle to locate a prize, drop a nickel, and hope the crane grabs the chosen toy and drops it down the chute. The best prizes were often out of reach or craftily glued down, so diggers were outlawed in some localities. Exhibit Supply Co. 1926.

ERIE DIGGER

The first digger is all mechanical, but offers no prize selection. The crane sweeps across the toys and stops at a different spot each time. The player continues to crank throughout the action. Erie Mfg. Corp. 1924.

CHICAGO DIGGER

Realism is attempted with colorful scenes of industrial sites or shipyards and the appropriate crane. With the lure of possible prizes in full view, diggers have been popular arcade games since their beginnings. Buckley Mfg. 1934.

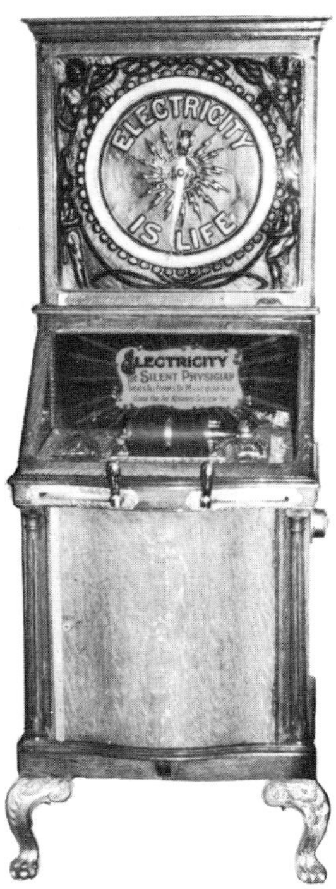

ELECTRICITY IS LIFE

It tingles, but it doesn't hurt. A mild shock is produced when the handles are grasped and moved apart to increase the current. The hand on the dial measures player stamina. Made by Mills Novelty Co. in 1904.

"BE HEALTHIER"

Originally electricity machines were considered therapeutic, but were later placed in arcades for competitive fun. The amount of current received is plainly shown for challengers trying to out endure each other. Roover Co. Pat. Feb. 16, 1909.

SPEAR THE DRAGON

Fantasy adds to the challenge. Endurance is visibly rewarded by seeing the warrior move across the bridge to spear the dragon and ring the bell. The hardy player is the dragon slayer. Made by Exhibit Supply Co. about 1927.

HIGH BALL

On this lung tester, for each 50 lbs. that the large nickel-plated ball passes, one of the nine bulbs lights up. The bell rings for a top score of 500 lbs. Mills Novelty Co. 1904.

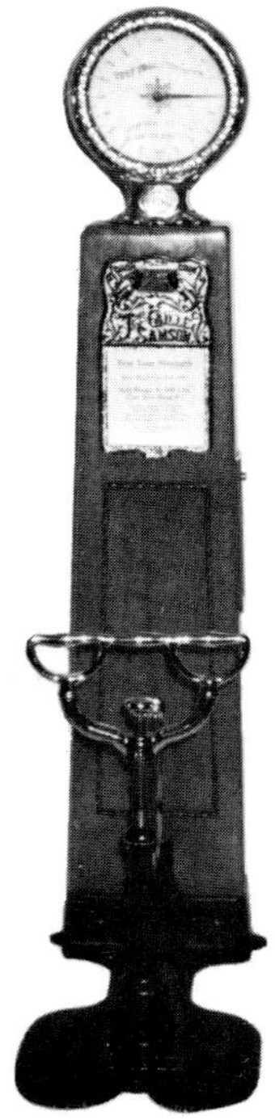

SAMSON LIFTER

How much can you lift? Strength testers like the above bring out the he-man qualities in the sporting young men who frequent arcades. Caille Bros. Pat. Feb. 28, 1905.

OWL LIFTER

The famous owl trademark of Mills Novelty Co. frames the well-muscled man pictured on the dial. The last score remains to dare the next contestant. 1904.

STRIKING CLOCK

Clock hands move around the dial as the handles are squeezed together. When they are released, the clock strikes the hour registered on its face. Exhibit Supply Co. 1925.

ROOVER STAMPER

See your name in print. After pressing out the letters with each turn of the dial, the printer will receive an embossed aluminum strip. Roover Mfg. Co. Pat. Nov. 19, 1901.

HARVARD STAMPER

This is the popular stamper for dispensing round souvenir tokens. Its successors are still found in arcades today. Harvard Automatic Machine Co. c1923.

TRUE HOROSCOPE

Select your reading from the month of your birth. The illuminated mystical scene features a papier-mâché swami and his two voluptuous attendants. Exhibit Supply Co. 1932.

THE ILLUSION

This bewitching machine uses mirrors to show the patron a transformation from his reflected image to a baboon. Another model shows "How You Will Be" —A Skeleton. Mills. 1904.

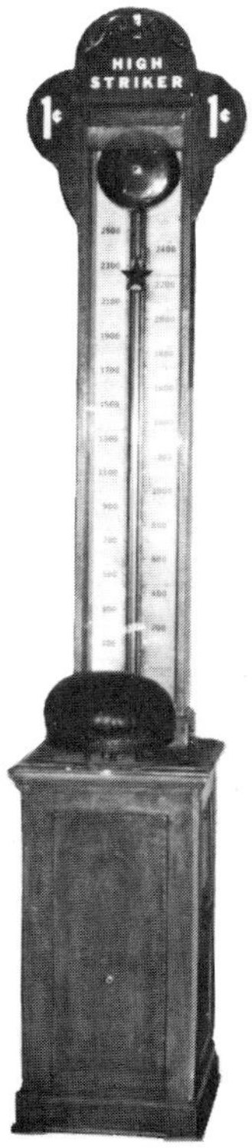

HIGH STRIKER

A powerful downward blow on the leather-covered horsehair cushion shoots the star up the post to ring the bell. Made by Gatter Novelty Co. c1918.

OLYMPIA PUNCHER

Grasping the handles and pressing against the cushion proves power of chest, and punching the cushion demonstrates the force of one's blow in Caille Brothers' two-way strength tester. Pat. Feb. 28, 1905.

DELIVER THE PUNCH

Emulate John L. Sullivan with no risk of a black eye. The colorful boxers pictured on the dial complete the illusion of a boxing arena. Mills Novelty Co. 1904.

CHAMPION

Keep punching at a penny a punch. Exhibit Supply Co. claims continued practice develops strong muscles. After each blow, the leather bag locks back into the wooden collar. 1925.

DUMB BELL LIFTER

Lifting the barbells moves the hand around the dial. On release, a bell records up to twelve gongs according to the power exerted. Automatic Novelty Co. c1925.

LIGHTHOUSE

The strength of a one-handed grip propels the lights to the top of the tower to ring the bell. Competition on strength testers often includes side bets. Globe Mfg. c1924.

HERCULES

Brain or brawn? The average grip of professions and trades are listed on the face. Caille Brothers' ornate cast-iron grip tester stands on a marbleized base. 1904.

TIGER PULL

Catch the tiger by the tail and tug. The harder the pull, the louder the roar. The cast-iron figure is painted in realistic colors and mounted on oak. Exhibit Supply. c1928.

PLAY FOOTBALL

The most popular of the Chester-Pollard sports games is still operated in many arcades. When the steel ball comes to rest in front of a member of your team, quickly push the handle down to propel his leg out and hopefully score a goal. A fumble gives the ball to your opponent. Game is three touchdowns. Made in 1924.

PLAY GOLF

Beware of sand traps. The man on the fairway is controlled by a single player adjusting the golfer's position with one handle and swinging his club with another knob. The score for this par three course is registered on a dial. Note how the lithographed scene gives a feeling of spaciousness. Chester-Pollard Amusement Co. 1928.

PLAY THE DERBY

Frenzied competition is encouraged by the playing operation. Each racer rapidly spins his wheel to propel his horse around the track. (A slip-clutch mechanism prevents overspin.) The winner is announced by a numbered flag popping up in the tower. Chester-Pollard Amusement Co. 1929.

1937 WORLD SERIES

An art deco veneered cabinet houses either Yankee Stadium or the Polo Grounds in the memorable 1937 Subway Series. The machine utilizes a complex system of weights and pulleys to keep score, and a small motor to automatically eject the ball from the cast-iron pitcher. This is a later version of the first model made in 1927. Rock-Ola Mfg. Co.

THE REAL TEN PIN

Win or lose, your penny buys a gumball. But the attraction is the challenge of bowling a strike by shooting the marble into the pins. Wires are attached to each tenpin for resetting the next frame. William Gent Mfg. 1928.

LADDIE 1¢ GOLF

The kilt-clad Scotsman plays his native game. One lever positions the golfer and another swings the club. A ball is provided for each of the three holes, with a five-point penalty for landing in a lake. United Engineering Works. c1929.

PLAY BASKET BALL

A gumball drops into the shooter's hands. If he makes a basket, the ball returns for another shot. If not, the gum is delivered out the front. Happily, the same penny buys a second try. Made by Junior Mfg. Co. 1929.

CHUCK-O-LUCK

How many thoroughbreds can you bring across the finish line? Flip the balls and aim for the win, place, or show baskets. The horses must jump the barrier to stay in the race or they drop out of play. D. Gottlieb & Co. 1931.

KICKER AND CATCHER

The right knob controls the kicker. The left knob moves the catcher trying to intercept the ball as it falls through the pins. A successful catch entitles the player to reuse the same ball for a higher score. Baker Novelty Co. c1935.

B & M BALL GUM

Try for a home run. Flip the baseball onto the diamond to knock over the tabs representing base hits. A gumball is dispensed in the front. Notice that the billboard behind the lithographed playing field advertises B&M Products. 1932.

THE WIZARD

Q: "Will I be famous?" A: "Yes! For your appetite." The wizard finds his answers in the flaming cauldron on the colorful cast-aluminum front. Mills Novelty Co. made an all-wooden version in 1904, and the above in 1919.

?INFORMATION?

Q: "How many speakeasies in New York City?" A: "The latest estimate is 60,000." Turn the knob to select a question, drop a penny, and open the mirrored window to read the reply. Information Vending Machine Co. 1930.

COMET

Q: "Would it be best for me to marry?" A: "You'll starve if you don't." If you don't like the fortune, you may like the gum. O.D. Jennings made their variation of a fortune-teller in 1929.

LINDY STRIKER

Will you reach Paris or be lost in mid-ocean? It all depends on how hard you hit the lever to move the aeroplane. "America's first patriotic game of aviation" also vends a gumball and was made by S. Madorsky Co. in 1929.

FINGER STRIKER

This counter-size cast-aluminum strength tester measures the power of one's finger flick. A mighty blow rings the bell and the penny is returned. Made by the Charles Fey & Co. and patented on July 5, 1932.

INDOOR STRIKER

Another combination game and gum vender, this piece challenges the player to flip rubber balls into the channels by striking the lever. Getting a high score is more a matter of luck than skill. In and Outdoor Games Co. 1930.

BOOP-A-DOOP

Many variations of this counter-top skill game offered popular inexpensive entertainment during the depression. An expert flip of the lever sends the metal balls spinning past the "out" hole to the high-scoring slots. Pace Mfg. Co. 1932.

PIKE'S PEAK

"It takes steady nerves, correct timing, and a sure hand" to climb all five balls to the top by rapidly flipping the metal staircase. This challenging game is also a gumball vender. Made by Groetchen Tool Co. in 1939.

ROCK-IT

Speed as well as skill is required. Try to move the ball through the maze in less than sixty seconds by rocking the platform. The needle on the dial times the action. Doraldina Corp. 1926.

POISON THIS RAT

"Pills" drop to the track at the top of the head which rolls from side to side as the knob is turned. The patriot works the poison into the mouth. Another version of this WWII favorite is "Kill the Jap." Groetchen Tool Co. 1941.

SHOOTOSCOPE

Drop a penny into the gun and shoot it at the target. Get a hit and a bell rings as your coin is returned. This game can take in money as fast as the player can shoot it. International Mutoscope Reel Co. 1926.

TARGET SKILL

A.B.T. Manufacturing Company became very successful by making many versions of this challenger. The above model automatically rewards the high-scoring player with a Sharp Shooter medal. Patented Sept. 1, 1925.

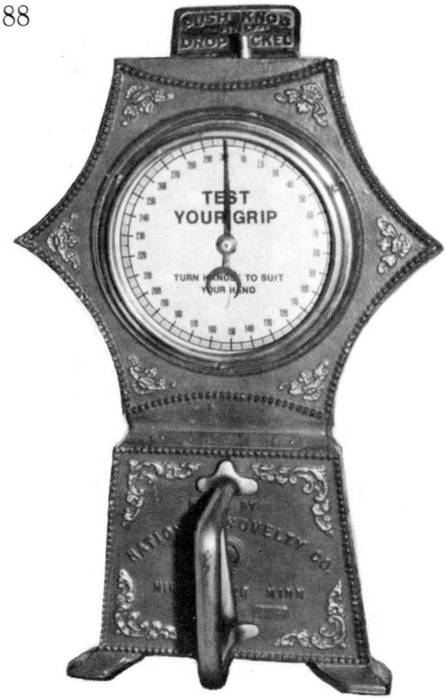

TEST YOUR GRIP

The eye-catching red and gold star shape attracts lots of nickels. The handle adjusts for the most comfortable grip to encourage repeated tries. Made by the National Novelty Co. about 1907.

TRY YOUR STRENGTH

Twist the knobs on each side of this fancy little cast-iron strength tester. The needle on the dial shows how strong you are. Its compact size makes this machine useable in many locations. Made by New England Mfg. Co. Pat. Jan. 22, 1901.

SPIROMETER

Testing lung capacity was apparently of therapeutic importance at the turn of the century. The medical endorsement on the front claims that using this machine will enlarge and strengthen the lungs. National Spirometer Co. c1899.

GRAPHOPHONE

Hearing mechanically reproduced music was novel enough to be worth five cents in 1899. When the spring motor is cranked, the needle travels across the spinning Edison cylinder to play a popular air or rousing march. Columbia Phonograph Co.

MOVING PICTURES

This early movie device was one of the first to employ a type of film strip. The crank operates a shutter mechanism which advances a strip of photos past the lens to give the illusion of moving pictures. Pat. Aug. 24, 1897.

SCULPTOSCOPE

"Whiting's travel system embraces the world." See the Tower of Pisa or the Taj Mahal in color and three dimension. A set of twenty different views costs a penny, and each machine contains four sets. American Novelty Co. 1910.

ELECTRICITY IS LIFE

By the turn of the century, electricity had begun to alter daily life, although its nature was only vaguely understood. The popular definition was "a power of nature and a manifestation of energy." Small wonder that devices were made for "therapeutic" as well as utilitarian purposes. This example by Midland Mfg. Co. was introduced around 1900.

AUTOMATIC THERAPY

Expensive doctor visits could be avoided by use of a coin-operated treatment machine. This variation claims to measure lung capacity and expansion in addition to administering the electric tonic. A high score on the lung test returns the patron's nickel. The National Spirometer Co. 1900.

THE U.A.M.C.

Most of these mechanical cure-alls were placed in general stores or pharmacies, and for a penny claimed relief for such assorted ailments as rheumatism, nervousness, and one-night jags. However, the designs are less scientific than ornamental. This polished oak machine has floral decals surrounding the dial. United Automatic Machine Co. c1905.

IMPERIAL ELECTRIC

Electricity machines employ dry cell batteries to generate a tickling current. By turning or pushing one knob, the "patient" controls the intensity of flow, which is indicated on the dial. This machine's dial is bright red and decorated with zodiac signs. The antique-coppered iron front is embellished with roses and maidens. Mills Novelty Co. 1902.

Pinball Machines

The evolution of today's familiar pinball machine followed a somewhat complicated process. Development did not proceed in a straight line, but like the pinball itself, ricocheted from point to point on its journey. Here are some of the important events in pinball history:

The beginnings stem from bagatelle, a popular game in the mid-nineteenth century. Patrons would shoot balls with cue sticks on a table-sized playfield. A modification of this game, called "Improvements in Bagatelle," was invented for home use and patented on May 30, 1871, by Montague Redgrave. The game is made of wood and nails, the size of its field smaller than that of today's pinball. Mr. Redgrave's patent includes a spring-plunger, swinging gates, bells, cups, and most important, an inclined playfield. These innovations earn him the title of the "Father of Pinball."

Parlor bagatelles continued to be popular home games in the 1870s and 1880s, and many variations were sold. Then, in the 1890s, coin-slots were added. One game in particular was patented on January 23, 1894, by C. P. Young of York, Pennsylvania. The playfield was similar to the earlier bagatelles, but the entire machine was actually a check-payout gambling device. It included features such as a glass-covered playfield, a backglass, and a payout mechanism for dispensing tokens for winning shots.

Another patent for a coin-operated pinball game was granted on October 14, 1902, to the Caille Brothers Company. Called the "Log Cabin," it was a cast-iron variation of other bagatelle games but used a lever instead of a spring-plunger to shoot balls. It, too, was intended as a trade stimulator for gambling but was more simply designed than Young's version and did not automatically pay out.

On May 7, 1907, a patent was granted to Harry Reed of Salem, Massachusetts, for the first battery-operated electric pinball. This non-coin-operated game apparatus had a playfield similar to Redgrave's, but it stood on legs and was twice the size. When the balls fell into the proper cups, electrical contacts caused bells to ring.

Oddly enough, in spite of these improvements, coin-operated pinball was not yet successful. Much later, around 1930, it was reintroduced by the In and Outdoor Games Company with a model called "Whoopee." This nickel-play pinball seems to be the first modern floor model game, but, because of its expensive selling price, it did not realize any commercial success.

Finally, the following year, "Bingo" and "Baffle Ball" were introduced. Bingo, made by the Bingo Novelty Company, was probably the earlier of the two, but it was Baffle Ball, a similar design manufactured by D. Gottlieb & Co., that became the first commercially successful pinball game. And when Ray Moloney could not get delivery on this popular game, he produced his own "Bally Hoo," and so began the now famous Bally Manufacturing Corp. These games were small, inexpensive countertop machines that provided cheap, competitive fun for the Depression-era players. Legs were added again within a short time.

Many manufacturers entered the market and gradually new features were added to keep the player's interest. The best features were quickly copied and included in new designs and so began the pattern for the evolution of pinball design.

A major development came about when Harry Williams, pinball inventor and later founder of the Williams Manufacturing Corp., saw some players trying to cheat one of his creations by hitting it, he went home and invented the first "TILT" mechanism. From then on, any player getting too rough with a machine would lose his next ball, and maybe the rest of the game.

In 1933, Bally introduced "Rocket," the first modern payout pinball machine to add gambling excitement. Competing for a high score could also pay off in cash, merchandise, or free games. (Unhappily, the addition of the gambling factor gave pinball a stigma that still lingers in many states today.)

As the public indicated its liking for pinball, the business continued to grow. Mechanical games reached their pinnacle with the "Jigsaw" and "World Series" models by Rock-Ola Manufacturing. Later, batteries were added to power solenoids for mechanical propulsion, and to ring bells and flash lights. Later still, the batteries were replaced by step-down transformers that converted 110 volts to lower voltage, making the machines safer and simpler to operate and maintain, and providing additional power for increasing the playing action.

As more complicated electrical features were added to the field, the backglass, originally an instruction sign, was a logical place to house extra works and attract the player. At first only a few inches high, illuminated backglasses continued to grow in size and in complexity by the addition of motif scenes, scoring mechanisms, and other novelty features.

In 1947, when Harry Mabs introduced flippers to pinball, a new era in pinball began. All previous models became obsolete as manufacturers rushed to redesign their playfields to accommodate this new feature. Soon after, the pop bumper and kicking rubber were added, making it the game we know today.

Pinball design continues to evolve and has recently entered the era of solid-state technology. From its beginnings pictured here to the present, pinball remains the most popular, entertaining coin-operated machine of all.

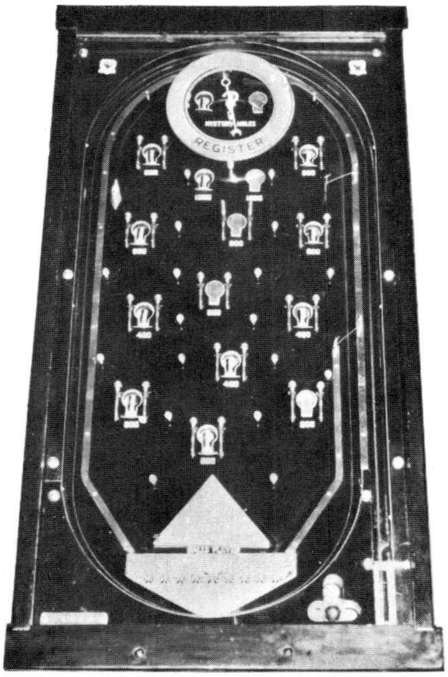

REGISTER

The name derives from its new feature, an automatic tally. "Toilet seat" lids close after accepting a ball, increasing the difficulty of the next shot. The sleek chrome, black bakelite, and wood cabinet is classic art deco. D. Gottlieb. 1932.

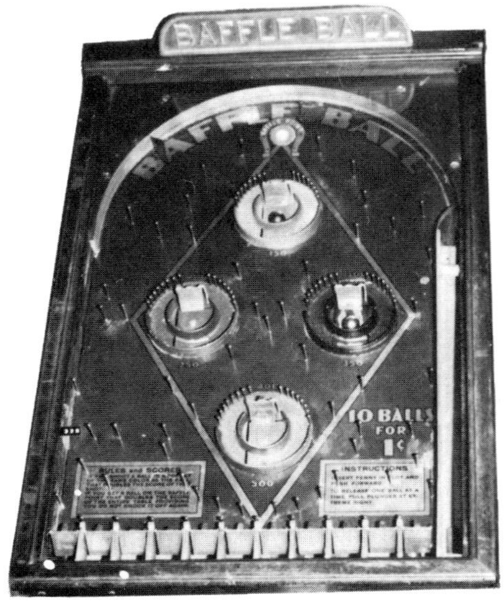

BAFFLEBALL

Elaborate scoring rules add to the fun on this first commercially successful pinball game. Matching the color of the ball with the color of the pocket doubles that point. A ball remaining on the Baffle Point doubles the score of the entire game. D. Gottlieb & Co. 1931.

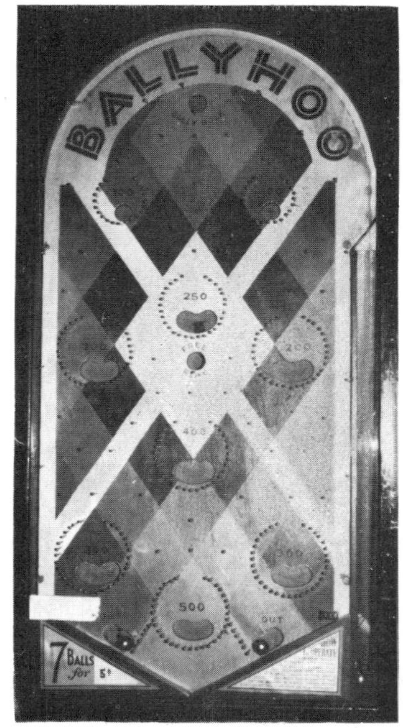

BALLY HOO

The now famous Bally Mfg. Co. named their first pinball machine after a popular satire magazine of the day. Besides the score-doubling Bally Hole, something different was added with the inclusion of a free play hole. 1931.

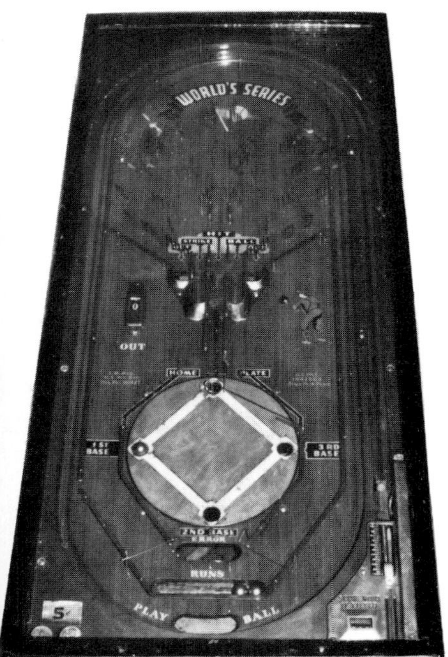

WORLD SERIES

Skill determines the length of the game. The player receives unlimited balls until hitting three outs. An all-mechanical apparatus revolves the diamond to advance balls from base to base on each hit, and automatically records the score. Rock-Ola Mfg. 1933.

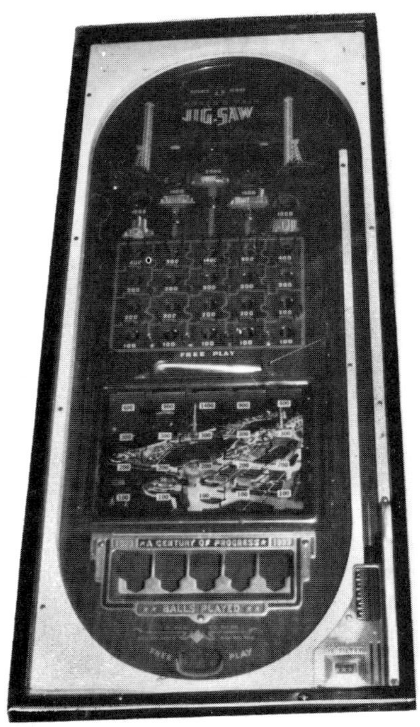

JIG-SAW

Try to complete the 1933 Chicago World's Fair puzzle. Shoot a ball into a bottom hole to set up a corresponding puzzle piece or land in a top hole to line up a whole column. This popular game increases enjoyment by visible reward. Rock-Ola. 1933.

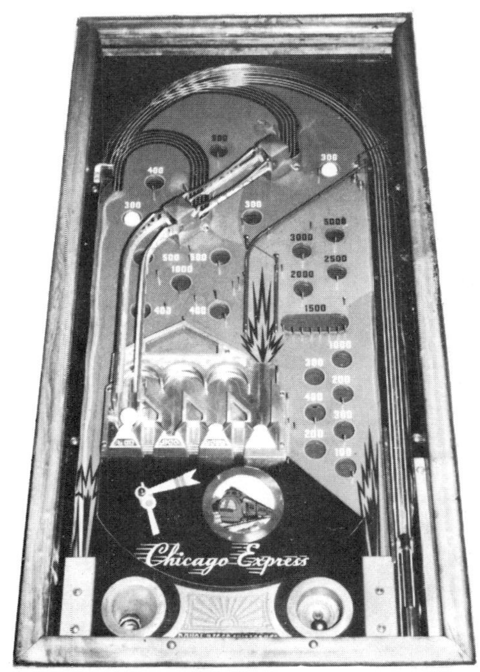

CHICAGO EXPRESS

Daval Mfg. Co. loaded this machine with all the new features available in 1934: ball elevator, kickers, a flashing light, and a bell. If the marble lifts onto the trestle, a signal light flashes and the marble is kicked into the high scoring area.

Restoration

In our age of sophisticated technologies, it is no longer possible to follow with our senses the inner working of today's machinery and appliances. While we may be impressed by their complexity, we are barely able to comprehend their construction. But when we enter the world of antique mechanical devices, free from riveted subassemblies and subminiaturized subcomponents, we can marvel at their relatively simple (but nevertheless ingenious) creations and enjoy tinkering with them and seeing how they work.

A machine that has somehow managed to survive in good to excellent original condition is the best example of its type. It is exciting and uncommon and worth more to a collector than even the finest similar machine that has been restored. The patina that develops from age gives the piece special qualities that cannot be achieved in a restoration project. Because most machines are found in less than presentable condition, however, some degree of restoration is necessary and acceptable, and will certainly increase its value. Although this type of work is not for everyone, to those who enjoy it each machine presents a challenge to return it to its functional and attractive state.

The role of the collector today is to preserve the relics of the past. Therefore, the objective of a restoration should be to re-create the appearance and performance of the machine as it was originally designed and shipped from the factory. For those who find themselves objecting to a completely new look, allowance can be made for "antiquing" a restoration by toning down such things as the gloss of paint, whiteness of paper, or brightness of nickel-plating. More often, these techniques are performed in order to blend in a new part or finish with the rest of the machine. A "balanced" restoration is aesthetically preferred. Customizing, modernizing, or personalizing a machine in any way not only subtracts from its authenticity but also lowers its value; it is not possible to restore a machine to "better than new."

There are two restorations for each machine: the mechanism and the case. A general guide to restoring follows. Fine points can only be learned through experience. For the restorer seeking quality results, consulting with other collectors is recommended. Another invaluable source of information is the reproductions of some early manufacturers' catalogs that provide specific information about metal and wood finishes, decals, marquees, etc. With care, extensive thought, common sense, and patience, a beautiful, working, and valuable machine will be the result.

The mechanism

Unjamming a mechanism: A nonworking machine should not be forced to operate because serious damage could result. The trouble spot must be carefully and patiently determined by trying to relieve spring-tension on jammed parts. Coins can get stuck, and are frequently the cause of jamming. Also, today's coins are not exactly the same thicknesses and weights as the coins the antique machines were designed to use. (For instance, many early payout machines will pay thirty-eight new nickels instead of forty old ones.)

Cleanliness: It is important that the mechanism is clean and free of dirt, sludge, cobwebs, and excess oil and grease. Washing with kerosene or other degreasers is effective, but remember to remove any paper parts first. Rust is very harmful and should be removed, either chemically or mechanically. After cleaning, the mechanism should be dry, or lightly coated with oil to impede further rusting.

Missing or broken parts: Transferring original parts from other machines to make a good machine is perfectly acceptable. Parts are interchangeable and many machines were repaired this way on location. If an original replacement part is not available, the next best solution is to faithfully reproduce an original. Sand casting and machining are two methods of copying parts. (When recasting a part, remember to consider the shrinkage factor.) Collectors are usually generous in making parts or loaning them for duplication. When parts are not available to copy, then any design that works reliably will suffice. Sometimes patents can be a helpful source of information. Many popularly needed items have already been reproduced for the collector's market, such as reel strips, handles, legs, and marquees. Broken metal can be welded and wood can be glued. Epoxy cement can be used when there is no stress involved.

Springs: Part and parcel of any mechanical device are various types of springs. Springs can sometimes be the source of very frustrating difficulties. Inspect them for proper fit and tension and replace only with close substitutes.

Lubrication: After the machine is cleaned and working, it is most important that it be properly lubricated in order to prevent wear and breakage, and to produce smooth operation. Except for spinning shafts, electrical contacts, and payout slides, which should be clean and dry, all other moving parts must be lubricated where they pivot or move against other parts. A No. 10 oil or sewing machine oil is used on small, lightweight parts and clockwork (spring) motors. Use a No. 20 oil on medium-sized parts such as bearings and camshafts. On heavy-wearing parts, use vaseline or cup grease. Wood should never be lubricated with oil because oil makes the wood swell. Use instead a mixture of graphite and vaseline. If a machine receives regular play, then lubrication should be done twice a year. Penetrating liquids or silicone sprays are good for loosening tight joints but are not satisfactory substitutes for grease and oil. And always be careful not to get lubricants on paper cards or reel strips, and do not overlubricate, because this attracts dirt and causes splattering.

The case

Cleanliness: The most appealing machine in either original or restored condition is one that looks clean and glossy. No one likes to buy from or play with a machine that is dirty. Before deciding on a major restoration, first try to retain the original elements by cleaning, waxing, and polishing. Having an original old, but clean and polished look can be more visually appealing and better in keeping with the machine's age than a completely new, refinished effect. Also, as more people rush to restore machines, the ones that remain in original condition will become more valued for their authenticity.

Wood: The Old World cabinetry work by expert craftsmen is an outstanding feature of many antique machines. Oak, a beautiful and easily restored hardwood, was the principal wood used. When refinishing, it is necessary for all mounted parts to be removed first (and a record kept of their locations). Stripping the wood should be done with a chemical remover that dissolves the finish and leaves in the original color. Paint removers can be used with caution, as they can bleach out the color. Commercial tub stripping is not recommended because it can damage the wood and loosen joints. Then steel-wooling, sanding, and mechanical repairs can be made, followed by an application of wood filler, to replace what has been removed. This is an important step that enhances the grain and yields professional results and one that was always done in the past (when labor was cheaper, of course). After this, stain can be applied if needed, and then the top coating. Two or three coats of clear gloss spar varnish is recommended, as it is the closest material to the original finishes used on most machines. Clear lacquer, also used then, can be used to achieve good results but requires more knowledgeable handling. Clear and orange shellac were also used on some early machines. For the sake of authenticity, polyurethane finish is not advised. For most machines, a medium to high gloss is standard. Some manufacturers' catalogs, however, list their finishes as "polished." This effect is achieved by properly hand rubbing the finish after it has thoroughly dried. The result will be a deep, rich, satin gloss.

Metal: Decorative metal parts should be bright and shiny. Aluminum, brass, and copper can be shined with a metal polish or on a buffing wheel with compound. Plated metals, mostly steel and iron, that are not worn, discolored, or peeling can be polished or buffed with care. When replating is necessary, it is important to match the original plate. Some early special plating techniques were striped antiqued copper and copper over nickel. Never use a wire brush for cleaning or polishing any aluminum or plated parts. When using a buffing wheel, care should be taken that fine detail in the part design is not erased. And when having a part replated, this is also good advice for the plater. (When using a buff, remember to wear safety goggles.)

Decals: Decals enhance the appearance and value of a machine and every effort should be made to save them. Even when under paint they can still be saved by carefully removing the paint. Some reproduction decals are available on the collector's market. Some decals have been duplicated by taking photographs of them and turning the photos into decals with transfer materials available in craft stores.

Paper: Paper parts, such as instruction cards, pay cards, and advertising cards can be duplicated from originals by many printing methods. New reel strips for three reel slots have been reproduced for collectors.

Paint: When repainting parts, try to match the old colors as closely as possible.

Porcelain: Glassware mending kits containing epoxy and glazing compounds with coloring pigments are handy for repairing small areas of damaged porcelain.

Glass: When replacing glass, try to duplicate the original thickness. When installing it, use a cushioning material, such as cardboard, between the glass and brackets in order to even out the pressure to prevent it from cracking. Crystal clear epoxy cement is effective for repairing cracks and chips. Specialists for beveling, mirroring, curving, and sandblasting can usually be found in large metropolitan areas.

Back doors: A machine is not complete if its mechanism is exposed. If a back door, cash box drawer, or similar part is missing, it should be constructed or copied from an original.

Locks and keys: If a machine is found locked with no key, every effort should be made to open it without damaging the case or drilling the lock. The locks have become valuable antiques as well and are often difficult to replace. A qualified locksmith should be easily able to open any antique lock. The final touch for any restoration is to have it fitted with a properly working lock.

NOTICE!

What's Wrong With Logan County?

Why are Slot machines flooding Indiana Lake?

Is someone paying off the County Officials?

What's wrong with the PROSECUTOR?

What's wrong with the JUDGE?

What's wrong with DEPUTIES?

Clark County was cleaned up recently and the Prosecutor is now serving time in the OHIO PENITENTIARY. How about a clean sweep for Logan County.

The kiddies are losing their NICKLES

THE COMMUNITY

The Index

• The location of each manufacturer, distributor, or operator is listed when known. A company may have had more than one address, however, if it moved, had branch offices, or merged with another company, so these are not final references.

• The height for each machine is included since the photographs cannot provide an accurate sense of dimension. Note the large variance in the sizes of machines.

• Pages listed in bold type indicate photographs.

DROP COIN
HERE